FIELD and LABORATORY INVESTIGATIONS in AGROECOLOGY

Third Edition

Stephen R. Gliessman

Professor Emeritus of Agroecology
University of California
Santa Cruz

CRC Press
Taylor & Francis Group
Boca Raton London New York

CRC Press is an imprint of the
Taylor & Francis Group, an informa business

CRC Press
Taylor & Francis Group
6000 Broken Sound Parkway NW, Suite 300
Boca Raton, FL 33487-2742

Printed on acid-free paper
Version Date: 20140619

International Standard Book Number-13: 978-1-4398-9571-9 (Paperback)

Library of Congress Cataloging-in-Publication Data

Gliessman, Stephen R.
    Field and laboratory investigations in agroecology / author, Stephen R. Gliessman. -- Third edition.
        pages cm
    Includes bibliographical references.
    ISBN 978-1-4398-9571-9 (alk. paper)
    1. Agricultural ecology--Fieldwork. 2. Agricultural ecology--Laboratory manuals. I. Title.

S589.7.G59 2014
577.5'5--dc23   2014021533

Visit the Taylor & Francis Web site at
http://www.taylorandfrancis.com

and the CRC Press Web site at
http://www.crcpress.com

Contents

Instructor's Preface v
Introduction ix
Author xv

Section I   Studies of Environmental Factors

1. Effect of Microclimate on Seed Germination 3

2. Light Transmission and the Vegetative Canopy 13

3. Soil Temperature 23

4. Soil Moisture Content 31

5. Soil Properties Analysis 39

6. Canopy Litterfall Analysis 51

7. Mulch System Comparison 57

8. Root System Response to Soil Type 67

Section II   Studies of Population Dynamics in Crop Systems

9. Intraspecific Interactions in a Crop Population 77

10. Management History and the Weed Seedbank 87

11. Comparison of Arthropod Populations 99

12. Census of Soil-Surface Fauna 107

Section III Studies of Interspecific Interactions in Cropping Communities

13. Bioassay for Allelopathic Potential 117

14. *Rhizobium* Nodulation in Legumes 133

15. Effects of Agroecosystem Diversity on Herbivore Activity 141

16. Herbivore Feeding Preferences 149

17. Effects of a Weedy Border on Insect Populations 157

Section IV Studies of Farm and Field Systems

18. Mapping Agroecosystem Biodiversity 167

19. Overyielding in an Intercrop System 173

20. Grazing Intensity and Net Primary Productivity 185

21. Effects of Trees in an Agroecosystem 193

Section V Food System Studies

22. On-Farm Energy Use 205

23. Farmer Interview 211

24. Local Food Market Analysis 217

Section VI Instructor's Appendix

Appendix A: Planning the Field and Laboratory Course 225

Section VII General Appendices

Appendix B: Measurement Equivalents 237

Appendix C: Material and Equipment Suppliers 239

Instructor's Preface

The purpose of this manual is to give students opportunities to gain direct hands-on field and laboratory experience with the concepts that make up the science of agroecology. Since agroecology is defined as the application of ecological concepts and principles to the design and management of sustainable food systems, such an experiential approach to teaching agroecology is essential.

This manual is specifically designed to be used for the field and lab component of a lecture-based classroom course in which *Agroecology: The Ecology of Sustainable Food Systems* serves as the main text. The 24 investigations are divided into five sections that roughly parallel the organization of the textbook, with each focusing on a distinct level of ecological complexity.

Investigations in Section I, Studies of Environmental Factors, are mainly autecological in nature. They deal with how an individual plant responds to the environment; how environmental factors, in specific agroecosystems, are measured and characterized; and how particular environmental factors affect individual plants.

In Section II, Studies of Population Dynamics in Crop Systems, investigations highlight how populations of organisms act in agroecosystems. They focus on what populations are present in a system, how populations change over time and respond to the environment, and how individuals within a population may interact.

The level of the community is the focus of Section III, Studies of Interspecific Interactions in Cropping Communities. Here, investigations explore the between-species interactions of the organisms that make up crop communities. These interactions include herbivory, allelopathy, and mutualisms.

Investigations in Section IV, Studies of Farm and Field Systems, look at system-level agroecology, examining either whole farms or systems within farm boundaries. These investigations touch on the complexity with which the farmer deals in managing agroecosystems.

Finally, in Section V, Food System Studies, investigations reach out beyond the individual farm to examine components of the food system at a local level, which impact all of the levels of analysis in the first four parts.

In addition to this parallel structure, *Field and Laboratory Investigations in Agroecology* provides the instructor with other tools for correlating the lecture and lab components of a course. Each investigation indicates the chapters in the text to which the investigation relates most directly, and an investigation's introductory section generally includes more specific references to textbook content. Further, a table at the end of this preface provides an overview of chapter–investigation correlation. Not all chapters are included in this table—Investigations 1 through 3 are general enough to have some connection to most investigations, and some later chapters cover topics (such as fire and genetics) that are not addressed in any investigations.

Despite the integration of this manual with *Agroecology: The Ecology of Sustainable Food Systems*, it can also be used without this text—as the core of a field and laboratory course, for example—or with another textbook in agroecology, agronomy, or ecology (in which case a copy of *Agroecology* will be handy for consultation).

Agroecology crosses traditional disciplinary boundaries, and the investigations in this manual reflect this "hybrid" nature of the science. In one sense, the investigations teach ecological concepts and principles as applied in agricultural practice; in another sense, they teach about agriculture from an ecological perspective. As a result, they can be used in an ecology or agronomy course just as profitably as in an agroecology course.

Because the investigations were selected and written with breadth and adaptability in mind, instructors from all types of climates and agricultural contexts should be able to find a coherent set of investigations to fulfill their needs and goals. Instructors should also be aware that the experimental methods used in each investigation do not exhaust the possible ways of investigating the questions at hand. Other procedures can be used, and instructors are encouraged to adapt these investigations to their needs.

The investigations are written for both students and instructors—students can use them as a source of step-by-step procedures and suggestions for data analysis and report writing; instructors will find them useful for planning, advance preparation, and supervision, and as starting points for designing their own investigations. Material specifically for instructors is included in the Instructor's Appendix; instructors should read this material before choosing investigations and planning the structure of the course, because it contains important information about materials and equipment, class-size assumptions, and so on.

As a teaching tool, the manual is focused on facilitating hands-on, experiential learning that involves close observation, creative interpretation, and constant questioning of findings. Because of this purpose, the goal of teaching methods of ecological investigation per se is secondary. The instructor who expects to find a detailed, in-depth instruction in statistical analysis, experimental design, and classic methods of ecological field study will be disappointed. Nonetheless, the investigations do teach the value of and need for rigor, consistency, and detail in the analysis of ecological systems. The importance of careful data interpretation and presentation is emphasized, as is the value of a clear, concise, and well-written research report. Simple statistical analysis, for data management and interpretation, is used throughout the manual, and students are guided through the steps of data analysis in the context of the particular investigations in which it is employed.

Correlations among Investigations and Selected Textbook Chapters	4 Light	5 Temperature	6 Humidity and Rainfall	8 Soil	9 Water in the Soil	11 Biotic Factors	12 Environmental Complex	14 Population Ecology	16 Species Interactions	17 Agroecosystem Diversity	18 Disturbance and Succession	19 Animals	20 Energetics	21 Landscape Diversity	23 Indicators of Sustainability	25 Community and Culture
Effect of microclimate on seed germination	✓	✓			✓		✓									
Light transmission and the vegetative canopy	✓															
Soil temperature		✓														
Soil moisture content			✓		✓											
Soil properties analysis				✓												
Canopy litterfall analysis				✓							✓					
Mulch system comparison		✓		✓	✓	✓										
Root system response to soil type				✓												
Intraspecific interactions in a crop population						✓		✓								
Management history and weed seed bank						✓		✓	✓							
Comparison of arthropod populations								✓	✓	✓						
Census of soil-surface fauna									✓	✓	✓					
Bioassay for allelopathic potential						✓			✓							
Rhizobium nodulation in legumes						✓			✓	✓						
Effects of agroecosystem diversity on herbivore activity								✓	✓	✓						
Herbivore feeding preferences								✓	✓	✓						
Effects of a weedy border on insect populations								✓		✓						
Mapping agroecosystem biodiversity										✓				✓		
Overyielding in an intercrop system								✓	✓	✓						
Grazing intensity and net primary productivity											✓	✓				
Effects of trees in an agroecosystem											✓					
On-farm energy use													✓			
Farmer interview															✓	✓
Local food market analysis																✓

Introduction

This manual is designed to teach ecological concepts and principles in an agricultural setting, to give you a much more in-depth and practical experience than you can glean from a textbook alone. You are challenged to view the fields or gardens where you work as ecological systems made up of complex interacting parts that respond to human manipulation and management. You are also encouraged to view agricultural systems as not just production systems, but rather as living, dynamic, evolving systems that must be adapted to the particular ecological and cultural conditions of each region of the world.

Investigations

This manual describes 24 different agroecological investigations. The description of each is designed to provide all the information you and your instructor will need to plan and carry out an experimental or comparative study, from background information to step-by-step procedures and suggestions for writing up the results. Most of the information in an investigation is directed to you, the investigator, but the section labeled Advance Preparation is meant to inform your instructor about what needs to be done before the investigation actually begins. You may or may not be called upon to take part in this preparation.

Keeping a Lab Notebook

Good science depends on careful record keeping. Because memory is always incomplete and very often faulty, everything you do in an investigation should be written down in a lab notebook. A well-kept lab notebook will help you work efficiently, stay organized, and have the necessary information at your fingertips when you need it.

Many kinds of information can be recorded in a lab notebook, including

- Hypotheses
- General notes and comments
- Data and observations
- Descriptions of how you set up experiments
- Diagrams of field and experimental setups
- Notes on variations or modifications in methodology
- Field notes
- Descriptions of progress in an investigation
- Flashes of insight

- Tentative explanations of results
- Conjectures and speculations
- Team agreements
- Calculations

As you can see from this list, a lab notebook serves at least three different purposes: it is (1) an objective record of scientific investigation, (2) a workspace, and (3) a record of your thinking and learning. To fulfill the latter purpose, you should try to make ongoing predictions and explanations of your results and observations. Construct hypotheses when applicable, and do as much interpretation of the data as you can.

Before you begin your lab notebook, decide how you will organize it. A suggested strategy is to organize it chronologically. Each day of the lab section, begin a new page with the date on top, and write a brief description of what you expect to do that day. Then record information related to that day's activities, even if it involves more than one investigation. If your lab notebook is a three-ring binder, you can organize it by subject or by investigation.

Using the Datasheets

Most of the investigations include one or more blank datasheets, which are designed to make it easier to record and analyze your data. The blank datasheets are found at the end of each investigation. Each datasheet is also shown filled in with sample data as an aid in understanding not only what the data should look like, but how they relate to the setup of the investigation and how they should be analyzed.

For many of the investigations, you will need more than one copy of a datasheet or set of datasheets. This occurs, for example, when you are comparing two different agroecosystems and one datasheet can accommodate the data from only one system. If there is a need for extra copies of any datasheets, this need is indicated in the Materials, Equipment, and Facilities section of the investigation. If you work on one of these investigations, be aware that you may be responsible for making the necessary photocopies.

Because of the frequent need for multiple copies of the datasheets, and because you may later need a blank copy of a particular datasheet, it is recommended that you do not write on the datasheets in this manual but instead use photocopies (or even handwritten facsimiles) for your data recording.

Working as a Team

The investigations are set up to be completed by teams. Each team is responsible for a certain portion of the investigation—a set of treatments, for example—and the investigation as a whole depends on every team doing its job well and not introducing errors or bias. The results obtained by each team are shared with the other teams to create a dataset larger and broader than one team alone can create. It is very important that all teams follow the same methodology, so that the data they obtain contain as little artificial variation as possible.

Good teamwork is essential for completing the team's tasks in a timely manner, minimizing error, and getting meaningful results. Team members must coordinate their activities, so that each person's responsibilities are clear cut and everyone is working toward a common goal. To achieve coordination, a team should discuss tasks and responsibilities at the beginning of each investigation and each class period, and reach an agreement on who will do what.

Using Statistical Analyses

None of the investigations require the use of complex statistical methods. The most advanced statistical tests are chi-square and *t*-tests, and most investigations involve only the calculation of means and sometimes standard deviations. When an investigation calls for a statistic to be calculated, the procedure for doing so is presented, and no advanced training in statistics is required.

However, you are encouraged to perform whatever statistical analyses you are capable of and feel are useful in the context of each investigation. In many of the investigations, for example, analyses of variance would be very useful (even though they are not part of the formal procedure). For this reason, a background in statistics is helpful.

Generating Random Numbers

Several of the investigations call for random sampling (of plants, of leaves, etc.). The goal is to obtain an unbiased sample of objects (such as 50 corn plants) that represent a larger group of those objects (such as a field of 1000 corn plants).

Random sampling in the field is best accomplished by (1) numbering each member of the large group (e.g., by row and column in a field), (2) preparing a list of random numbers, and (3) taking as the sample each member of the group whose number corresponds to one of the random numbers.

To prepare an appropriate list of random numbers for random sampling, you need to know two things: (1) the quantity of random numbers needed (this figure, which we will call Q, corresponds to the size of the sample) and (2) the approximate size of the large group being sampled (the universe). This latter piece of information is very important. If the universe has only 100 members, you do not want random numbers larger than 100 because they cannot yield any selections. Conversely, if the random numbers only range from 1 to 50, none of the members of the universe with a number higher than 50 can be selected, and the sample cannot be truly random. Ideally, then, the list of random numbers should range from 1 to a maximum magnitude (M) equal to the size of the universe. More precisely, each number between 1 and M should have an equal chance of being chosen in the process of generating the list of random numbers.

There are two basic methods for generating a list of random numbers of quantity Q and maximum magnitude M.

Method A—using a table of random numbers:

(a) If M is 9 or less, arbitrarily choose any location in a table of random numbers and read individual digits in sequence down or across, keeping those between 1 and M and throwing out those greater than M, until a quantity of random numbers equal to Q has been collected.

(b) If M is between 10 and 99, the procedure described in (a) is followed, except two digits at a time are read. If M is much less than 99, many numbers may have to be thrown away before Q is reached.

(c) If M is between 100 and 999, the procedure described in (a) is followed, except three digits at a time are read. If M is much less than 999, many numbers may have to be thrown away before Q is reached.

Method B—using an electronic random number generator (on a calculator or computer):

(a) Generate random numbers one at a time.

(b) Move the decimal point of each number the appropriate distance to obtain a number with the same number of places as $M - 1$.

(c) Ignore all numbers greater than M and record all numbers equal to or less than M until a quantity of random numbers equal to Q has been collected.

Writing Lab Reports

Besides serving as a basis for your instructor's evaluation of your performance, a lab report helps you understand the results of your investigation and allows you to communicate the results to others. In these respects, a lab report is very much like a scientific paper and should be treated as one. A person unfamiliar with the investigation should be able to understand from your report what you did, how and why you did it, what results you obtained, what you think the results mean, and what insights you have into applying the results.

The investigations in this manual are written to leave the interpretation, analysis, communication, and application of the results up to you (or you and your team, in the case of team reports). You may find that you can take the results in a variety of different directions, all equally valid. The investigations offer suggestions for what to include in your reports, but the results you get and your own unique approach to interpreting them will guide how you construct your reports.

Report Format

Scientific papers follow a basic format designed to present information clearly and in a logical order. The following chart lists the section headings usually included in a scientific paper and describes what each section typically contains.

Title	Fifteen or fewer words describing fully but concisely what the investigation was about.
Abstract	A one-paragraph summary of the report succinctly describing the objectives of the study, the methods used and measurements made, the results, and the conclusions.
Acknowledgments	A good place for explaining who in your group did what, and for noting any help you received.
Introduction	A short section describing the purpose of the investigation or what was being tested or explored—often includes some background information helping the reader understand the context and significance of the investigation.
Materials and Methods	A short description of what you did and how, including variations from the written procedure, and the statistical tests you used. As long as you reference the lab manual procedure, you need not repeat it.
Results	A presentation of the data (not an interpretation of them), accompanied by clearly labeled tables and/or graphs. Indicate the statistical significance of individual results, if appropriate; explain the meaning of the graphs and tables; and highlight trends, patterns, and significant results.
Discussion and Conclusions	An interpretation of the results, focusing on what they mean. Indicate if the objectives were met and whether the hypothesis was refuted or supported; speculate about mechanisms of action and cause-and-effect relationships; critically analyze your methods and speculate on possible sources of error; acknowledge your assumptions and the limitations of the study; make general conclusions; offer ideas for further study or application of the results; and explain what the results indicate about the importance of using an ecological basis for designing and managing agroecosystems.
Literature Cited	List any sources cited in the report.

Scientific Nomenclature

The Latin names of organisms should be given when appropriate. A rule of thumb: identify crops with their common names (including the variety or cultivar) and identify weeds, disease organisms, insects, other non-farm animals, and native plants with their Latin names. If giving both a Latin name and a common name, be consistent about which you list first, and put the second one in parentheses.

Latin names of organisms consist of two words: the genus name comes first and is capitalized, and the specific name is second and is not capitalized. Both words are either italicized or underlined (e.g., *Avena fatua*). If a particular Latin name is used more than once, each subsequent mention can use a one-letter abbreviation of the genus name (e.g., *A. fatua*).

Constructing and Labeling Graphs and Tables

Creating clear, easily understandable graphs and tables is crucial to communicating your results. Here are some guidelines to follow. You may want to study published scientific papers for examples.

- Present data in the most appropriate form. A table is adequate for a summary of data; but when you want to compare results or show trends and patterns, a graph is usually better.
- Use the appropriate type of graph. The type of data you want to present, as well as your purpose in presenting it, will determine which type of graph is best. Much of the data you will present is suited to simple bar graphs, but sometimes you will need to use stacked bar graphs or other complex bar graphs, line graphs, or even pie charts.
- Choose titles that describe exactly what a table or graph contains.
- Label graphs and tables clearly. Everything should be labeled: columns and rows in tables, and bars, lines, horizontal axes, vertical axes, and units of measure in graphs. Labels should be concise and precise, leaving no room for confusion or ambiguity. If you performed a test of statistical significance (e.g., a t-test or chi-square test) on the data you are presenting, indicate which results are significantly different from each other and at what level of significance or confidence (e.g., $p \geq .05$).
- Always explain the meaning or content of graphs and tables in your discussion; do not let them stand on their own.

Author

Stephen R. Gliessman earned a BA in botany, an MA in biology, and a PhD in plant ecology from the University of California, Santa Barbara, and has accumulated more than 40 years of teaching, research, and production experience in the field of agroecology. His international experiences in tropical and temperate agriculture, small-farm and large-farm systems, traditional and conventional farm management, hands-on and academic activities, nonprofit and business employment, and organic and synthetic chemical farming approaches have provided a unique combination of experiences and perspectives to incorporate into this text. He was the founding director of the University of California, Santa Cruz (UCSC), Agroecology Program, one of the first formal agroecology programs in the world, and was the Alfred and Ruth Heller Professor of Agroecology in the Department of Environmental Studies at UCSC until his retirement in 2012. He is the cofounder of the nonprofit Community Agroecology Network (CAN) and serves as the president of its board of directors. He is the editor of the international journal *Agroecology and Sustainable Food Systems*.

Studies of Environmental Factors

Effect of Microclimate on Seed Germination

Background

Microclimate is defined as the environmental conditions in the immediate vicinity of an organism. These conditions, which include moisture, temperature, and light, interact with each other and together affect the organism's growth and functioning.

The effect of microclimate on an organism is particularly pronounced in the case of a seed in the soil waiting to germinate. Seeds germinate in response to a precise set of conditions in their immediate environment (Figure 12.2 of *Agroecology: The Ecology of Sustainable Food Systems*). These conditions are specific to each plant species. The locality at the scale of the seed that provides the microclimatic conditions necessary for germination has been termed the "safe site." The conditions of the safe site must endure until the seedling becomes independent of the original seed reserves.

Textbook Correlation

Investigation 3: The Plant
Investigation 4: Light (Other Forms of Response to Light: Germination)
Investigation 5: Temperature (Responses of Plants to Temperature)
Investigation 9: Water in the Soil
Investigation 12: The Environmental Complex (Complexity of Interaction)

Synopsis

Seeds of two indicator species are placed in petri dishes (germination chambers) containing clean, sterile, screened sand. Conditions immediately surrounding the germinating seeds are maintained at a range of specific levels for moisture, light, and temperature. Observations of germination rates are made, and these results are correlated with the actual microclimatic conditions of each growth chamber.

Objectives

- Study the germination process.
- Control and vary microclimatic conditions.
- Correlate variation in germination response to microclimate variation.

Procedure Summary and Timeline

Prior to week 1

- Select test species, obtain seeds, and collect materials.

Week 1

- Set up germination chambers.

Week 2

- Observe and record numbers of seeds germinating.

After week 2

- Analyze data and write up results.

Timing Factors

This investigation is independent of outdoor conditions and can be performed any time of the year. It lends itself to rapid setup and takedown and can be completed in only 2 weeks.

Materials, Equipment, and Facilities

Twenty-four 9 cm diameter petri dishes (glass or plastic) per team (96 total)

5 kg of clean, sterile sand (run through a 30-mesh screen)

Parafilm for sealing dishes

10 mL pipettes

Distilled water

9 cm #1 filter paper discs

Tweezers

Tape (or labels) and marking pens

Approximately 500 seeds each of two test species

Climate-controlled incubators (or appropriate substitutes, which may include fluorescent lights, 24 h timers, portable heaters, and maximum–minimum thermometers)

Light meter (optional)

Two copies of the Germination Microclimate Datasheet, per student

Advance Preparation

- Select two test species. Annual crops, such as corn, beans, cucumbers, and lettuce, are good candidates since they have been selected for their rapid germination and uniformity of response, but they may have specific safe-site requirements. Less genetically manipulated annual plants used in agriculture, such as oats, barley, fava beans, or clover, would also work well. They may have broader tolerance of conditions. Finally, weed seed, if it can be obtained, would be interesting to test, since weeds have been selected for more variable and disturbed situations.

- If climate-controlled incubators are not available, develop a plan for how to vary and control the light and temperature conditions for the different microclimate tests and acquire the necessary equipment. For the light variation test, for example, it will be necessary to provide a dark location, a location with light 24 h a day, and a location with alternating periods of light and dark. See "Procedure" section for more information on what is required.

- If necessary, screen, wash, and sterilize the sand. Screened and washed sand can be sterilized by spreading it on trays and heating for 24 h at 105°C.
- Photocopy the appropriate number of datasheets.

Ongoing Maintenance

If climate-controlled incubators are not available, and ways of controlling environmental conditions are improvised, it will be necessary to check daily the different locations where the petri dishes are kept, monitor conditions, and alter the conditions if necessary.

Investigation Teams

Form four teams, with three to five members per team. Each team will be responsible for testing the responses of the two test species to variations in one microclimate condition. See the following logistics map.

Team 1	Team 2	Team 3	Team 4
Test response to moisture variation	Test response to temperature variation	Test response to light regime variation	Test response to seed position variation
3 moisture treatments per species × 4 replicates per treatment × 2 species = 24 germination chambers	3 temperature treatments per species × 4 replicates per treatment × 2 species = 24 germination chambers	3 light regime treatments per species × 4 replicates per treatment × 2 species = 24 germination chambers	3 seed position treatments per species × 4 replicates per treatment × 2 species = 24 germination chambers

Procedure

To test the effects of one microclimate variable at a time, the other variables must be held constant. For this reason, we have defined a set of standard conditions for each of the microclimate variables. These are as follows: 10 mL of water, 25°C, constant darkness, and the seed embedded at least partially in the sand. For each test, three of these conditions hold and the other is varied. The way this works is summarized in Table 1.1.

Each test of a variable requires a somewhat different experimental setup. These setups are described in the following; follow the setup procedure to which your team is assigned. After your team records its results, share them with the other teams.

TABLE 1.1
Summary of Microclimate Test Conditions

	Test Variations			
	Moisture Variation	Temperature Variation	Light Variation	Seed Position Variation
Moisture	8 mL; 10 mL; 12 mL	10 mL	10 mL	10 mL
Temperature	25°C	15°C; 25°C; 35°C	25°C	25°C
Light	Constant dark	Constant dark	Constant dark; alternating; constant light	Constant dark
Seed position	Embedded in the sand	Embedded in the sand	Embedded in the sand	Uncovered; embedded; under paper

Moisture Variation Setup

1. Prepare the germination chambers.
   a. Obtain 24 petri dishes.
   b. Place 45 g of sand into each dish and level the surface.
   c. Affix a label or tape to the lid of each dish and record for each dish the team name, date, seed species, condition being varied (moisture), treatment, and replicate number. The following matrix shows the resulting identity of each of the 24 dishes:

	8 mL Treatment	10 mL Treatment	12 mL Treatment
Species A	Replicates 1, 2, 3, 4	Replicates 1, 2, 3, 4	Replicates 1, 2, 3, 4
Species B	Replicates 1, 2, 3, 4	Replicates 1, 2, 3, 4	Replicates 1, 2, 3, 4

2. Place the seeds in the dishes.
   a. Arrange 10 seeds of the appropriate species in a circle in each dish, equidistant from each other. Each individual seed should be approximately 1 cm from the edge of the dish. If the point of emergence of the hypocotyl (root) can be determined, place that end inward.
   b. Press each seed into the sand with tweezers so that at least half of the seed is embedded in the sand.
3. Irrigate the dishes.
   a. Pipette exactly 8 mL of distilled water into each of the eight dishes labeled with the 8 mL treatment.
   b. Allow the water to discharge slowly from the pipette. Distribute the water evenly and be careful not to disturb the seeds.
   c. Using the same technique, pipette exactly 10 mL of water into each of the eight dishes labeled with the 10 mL treatment.
   d. Pipette exactly 12 mL of water into each of the eight remaining dishes, labeled with the 12 mL treatment.
4. Place lids on the dishes and seal them with parafilm to prevent drying.
5. Place all 24 dishes in a dark location at 25°C. If climate-control growth chambers are not available, find the best location possible and place a maximum–minimum thermometer in the same location to record the temperature and its variation.

Temperature Variation Setup

1. Prepare the germination chambers.
   a. Obtain 24 petri dishes.
   b. Place 45 g of sand into each dish and level the surface.
   c. Affix a label or tape to the lid of each dish and record for each dish the team name, date, seed species, condition being varied (temperature), treatment, and replicate number. The following matrix shows the resulting identity of each of the 24 dishes:

	15°C Treatment	25°C Treatment	35°C Treatment
Species A	Replicates 1, 2, 3, 4	Replicates 1, 2, 3, 4	Replicates 1, 2, 3, 4
Species B	Replicates 1, 2, 3, 4	Replicates 1, 2, 3, 4	Replicates 1, 2, 3, 4

2. Place the seeds in the dishes.

   a. Arrange 10 seeds of the appropriate species in a circle in each dish, equidistant from each other. Each individual seed should be approximately 1 cm from the edge of the dish. If the point of emergence of the hypocotyl (root) can be determined, place that end inward.

   b. Press each seed into the sand with tweezers so that at least half of the seed is embedded in the sand.

3. Irrigate the dishes.

   a. Pipette exactly 10 mL of distilled water into each dish.

   b. Allow the water to discharge slowly from the pipette. Distribute the water evenly and be careful not to disturb the seeds.

4. Place lids on the dishes and seal them with parafilm to prevent drying.

5. Place the dishes in three different locations that vary only in temperature. All locations should be dark.

   a. Place the eight dishes labeled with the 15°C treatment in a climate-controlled growth chamber set at 15°C. If such a chamber is not available, find a cool location at approximately 15°C and place a maximum–minimum thermometer at the site to record variations in the temperature.

   b. Place the eight dishes labeled with the 25°C treatment in a climate-controlled growth chamber set at 25°C. If such a chamber is not available, find a room-temperature location at approximately 25°C and place a maximum–minimum thermometer at the site to record variations in the temperature.

   c. Place the eight dishes labeled with the 35°C treatment in a climate-controlled growth chamber set at 35°C. If such a chamber is not available, find a warm location at approximately 35°C and place a maximum–minimum thermometer at the site to record variations in the temperature.

6. If climate-controlled growth chambers are not used, monitor the temperature of each location and make adjustments as necessary. The maximum–minimum thermometers will allow you to determine exactly how much the temperature varied from the intended temperature during the study period.

Light Variation Setup

1. Prepare the germination chambers.

   a. Obtain 24 petri dishes.

   b. Place 45 g of sand into each dish and level the surface.

   c. Affix a label or tape to the lid of each dish and record for each dish the team name, date, seed species, condition being varied (light), treatment, and replicate number. The following matrix shows the resulting identity of each of the 24 dishes:

	Dark Treatment	Alternating Treatment	Light Treatment
Species A	Replicates 1, 2, 3, 4	Replicates 1, 2, 3, 4	Replicates 1, 2, 3, 4
Species B	Replicates 1, 2, 3, 4	Replicates 1, 2, 3, 4	Replicates 1, 2, 3, 4

2. Place the seeds in the dishes.

   a. Arrange 10 seeds of the appropriate species in a circle in each dish, equidistant from each other. Each individual seed should be approximately 1 cm from the edge of the dish. If the point of emergence of the hypocotyl (root) can be determined, place that end inward.

   b. Press each seed into the sand with tweezers so that at least half of the seed is embedded in the sand.

3. Irrigate the dishes.
   a. Pipette exactly 10 mL of distilled water into each dish.
   b. Allow the water to discharge slowly from the pipette. Distribute the water evenly and be careful not to disturb the seeds.
4. Place lids on the dishes and seal them with parafilm to prevent drying.
5. Place the dishes in three different locations that vary only in the light they receive. All locations should be at the same temperature.
   a. Place the eight dishes labeled with the constant dark treatment in a dark climate-controlled growth chamber set at 25°C. If such a chamber is not available, find a dark location at approximately 25°C and place a maximum–minimum thermometer at the site to record variations in the temperature.
   b. Place the eight dishes labeled with the alternating light and dark treatment in a climate-controlled growth chamber set at 25°C. Program the chamber to cast light on the dishes for 12 h and then be dark for 12 h. If such a chamber is not available, find a dark location at approximately 25°C and set up a light attached to a timer. Set the timer to turn on the light sometime in the morning and go off 12 h later. Place a maximum–minimum thermometer at the site to record variations in the temperature.
   c. Place the eight dishes labeled with the constant light treatment in a climate-controlled growth chamber set to cast light 24 h a day and remain at 25°C. If such a chamber is not available, find a location at approximately 25°C and set up a light to be on 24 h a day. Place a maximum–minimum thermometer at the site to record variations in the temperature.
6. If desired, use a light meter to measure the light intensity for each light treatment.

Seed Position Variation Setup

1. Prepare the germination chambers.
   a. Obtain 24 petri dishes.
   b. Place 45 g of sand into each dish and level the surface of the sand.
   c. Affix a label or tape to the lid of each dish and record for each dish the team name, date, seed species, condition being varied (seed position), treatment, and replicate number. The following matrix shows the resulting identity of each of the 24 dishes:

	Placed on Surface Treatment	Embedded-in-the-Sand Treatment	Covered-with-Paper Treatment
Species A	Replicates 1, 2, 3, 4	Replicates 1, 2, 3, 4	Replicates 1, 2, 3, 4
Species B	Replicates 1, 2, 3, 4	Replicates 1, 2, 3, 4	Replicates 1, 2, 3, 4

2. Place the seeds in the dishes.
   a. Arrange ten seeds of the appropriate species in a circle in each dish, equidistant from each other. Each individual seed should be approximately 1 cm from the edge of the dish. If the point of emergence of the hypocotyl (root) can be determined, place that end inward.
   b. For each of the eight dishes labeled with the embedded-in-the-sand treatment, press each seed into the sand with tweezers so that at least half the seed is embedded in the sand.
   c. Leave the seeds in the other 16 dishes sitting on the sand surface.
3. Irrigate the dishes.
   a. Pipette exactly 10 mL of distilled water into each dish.

b. Allow the water to discharge slowly from the pipette. Distribute the water evenly and be careful not to disturb the seeds.

c. For each of the eight dishes labeled with the covered-with-filter-paper treatment, cover the seeds with a 9 cm disc of filter paper moistened with distilled water.

4. Place the lids on the dishes and seal them with parafilm to prevent drying.

5. Place all 24 dishes in a dark location at 25°C. If climate-control growth chambers are not available, find the best location possible and place a maximum–minimum thermometer in the same location to record the temperature and its variation.

Data Collection

Data collection (observation of germination rates) will normally take place 1 week after setting up the petri dishes.

1. Open each dish. Count the number of germinated seeds (if the seeds were buried, it may be necessary to uncover them to check for germination). For a seed to count as germinated, it must have an emerged root at least 1 mm in length.

2. Record the number of germinated seeds for each dish on the Germination Microclimate Datasheet. (Make a photocopy and use a different datasheet for each test species.)

Data Analysis

1. Calculate seed germination totals for each treatment. Calculate the percentage of germinated seeds (total/40 × 100) for each treatment. Record these figures on the datasheet.

2. Calculate a mean for each treatment (the average number of germinated seeds in each dish) and enter this figure in the datasheet.

3. Calculate a standard deviation (s) for each treatment. (Standard deviation is a measure of variation around the mean for set of observations.) If the data are entered in an electronic spreadsheet, the standard deviation can be calculated automatically with the standard deviation function in the software; a pocket calculator with the same kind of function can also be used. Otherwise, follow the procedure described in the following:

a. For each set of four data points, calculate the "sum of squares." Each observation (x_i the number of germinated seeds in a dish) is subtracted from the mean ($\bar{x}$) for the set of four observations, the result is squared, and then all four squared results are added together. This process is easily done using a handwritten chart modeled on the following.

Observation x_i	Mean $\bar{x}$	Difference $x_i - \bar{x}$	Square $\left(x_i - \bar{x}\right)^2$
		Sum of squares = $\sum \left(x_i - \bar{x}\right)^2$	

b. Use the following formula to calculate the standard deviation (s) for each set of data:

$$s = \sqrt{\frac{\text{Sum of squares}}{n-1}}$$

4. Display, post, or otherwise publish your team's data so that each team can fill in all of the data on the datasheet and use it in their write-ups.

Write-Up

The following are some suggestions for writing up the results of the investigation:

- Summarize the results, using bar graphs that allow comparison of germination means and/or germination percentages among treatments and between test species. The bar graphs of means can include error bars showing a range one standard deviation higher than each mean and one standard deviation lower. Identify the treatments showing statistically significant differences from other treatments.
- Discuss patterns and correlations between microclimate conditions and seed germination success. How do light, moisture, and position appear to affect the germination of seeds? Which factor seems to be the most important for germination? The least important?
- Discuss conclusions and propose questions for further study. How might the experimental design be altered to allow you to investigate the new questions?

Reports can also include statistical testing of the results (i.e., a chi-square test) to determine if the differences among germination rates are significant. You should attempt this analysis only if you have a background in statistics.

Variations and Further Study

1. Adapt the investigation to the field. Sow seeds of the same species in a variety of locations with different conditions, monitor those conditions over a week's time, and then gather data on germination rates.
2. Increase the range of variation for each variable. If time, space, and materials allow, add both a higher level and a lower level to each treatment set, for a total of five levels per set. For the moisture variable, for example, include treatments of 6 and 14 mL in addition to those of 8, 10, and 12 mL.
3. Rather than test the impact of a range of microclimatic conditions on two plant species, keep the microclimatic conditions constant and test a large range of species. This approach will shed light on genetic variation in safe-site requirements.
4. Test the impact of the orientation of the seed (e.g., hypocotyl up, down, and horizontal) on germination success (Figure 1.1).

Germination Microclimate Datasheet								
Date: November 23, 2014								
Test Species: Pole Bean								

Condition Varied	Treatment	Number of Germinated Seeds							% Germination
		Replicates				Total	Mean	Standard Deviation	
		1	2	3	4				
Moisture	8 mL	9	8	9	8	34	8.5	0.58	85
	10 mL	10	10	9	10	39	9.75	0.50	98
	12 mL	6	7	9	8	30	7.5	1.29	75
Temperature	15°C	4	1	2	1	8	2	1.41	20
	25°C	10	6	10	9	36	8.75	1.89	90
	35°C	10	10	10	9	39	9.75	0.50	98
Light	Constant dark	9	9	10	9	37	9.25	0.50	93
	Alternating light and dark	10	10	7	9	36	9	1.41	90
	Constant light	8	8	10	10	36	9	1.16	90
Seed position	Uncovered	5	6	7	4	22	5.5	1.29	55
	Embedded in the sand	9	10	10	9	38	9.5	0.58	95
	Under filter paper	9	10	9	10	38	9.5	0.58	95

Figure 1.1
Example of a Completed Germination Microclimate Datasheet.

		Number of Germinated Seeds							
Condition Varied	Treatment	Replicates				Total	Mean	Standard Deviation	% Germination
		1	2	3	4				
Moisture	8 mL								
	10 mL								
	12 mL								
Temperature	15°C								
	25°C								
	35°C								
Light	Constant dark								
	Alternating light and dark								
	Constant light								
Seed position	Uncovered								
	Embedded in the sand								
	Under filter paper								

Germination Microclimate Datasheet

Date:

Test Species:

2

Light Transmission and the Vegetative Canopy

Background

An agroecosystem is in essence an ecosystem designed to capture the energy in sunlight and use it to fix atmospheric carbon for the production of edible biomass. An important aspect of agroecosystem function, therefore, is the efficiency with which the capture of light energy takes place. How much of the solar energy reaching the system strikes leaf surfaces? How much of this energy is absorbed and how much is transmitted and reflected? The efficiency of light capture, in turn, is in large part a function of the structure of the vegetative canopy. Does the canopy allow a significant portion of the incident light to reach the ground, where it can only promote the growth of weeds? Does an upper layer of foliage from one species receive more light than it can capture, reducing the amount of light received by the foliage of other crop species given in the following?

Answering these kinds of questions involves two basic avenues of investigation: analyzing the structure of the canopy as it relates to light capture and analyzing the internal light environment—how light penetrates the canopy and is progressively absorbed down through the vertical structure of the system (Chapter 4 of *Agroecology: The Ecology of Sustainable Food Systems*). With a better understanding of the relationship between agroecosystem structure and efficiency of light capture, we have an important basis for maximizing productivity in the design of agroecosystems (Chapter 4 of *Agroecology: The Ecology of Sustainable Food Systems*).

Textbook Correlation

Investigation 3: The Plant
Investigation 4: Light

Synopsis

Three existing cropping systems are analyzed to determine for each (1) the number of layers of leaves making up the canopy, as given by the leaf area index (LAI), and (2) the amount of light transmitted through the canopy to the ground. These measurements are used to characterize and interrelate the light environment and canopy structure of each system and provide useful information on how each system absorbs and captures sunlight.

Objectives

- Investigate how light is distributed in and absorbed by an agroecosystem.
- Get an impression of how agroecosystems vary in their canopy structures and internal light environments.
- Compare the effects of different cropping structures on the internal light environment.
- Learn a technique for determining the LAI of a cropping system.
- Learn techniques for measuring light quality, distribution, and capture.
- Consider the effects of the internal light environment on crop development.

Procedure Summary and Timeline

Week 1

- Collect data on the canopy structure and light transmission of the three different cropping systems.

After week 1

- Analyze data and write up results.

Timing Factors

This investigation can be performed at almost any stage in the development of the cropping systems being examined. The measurements of light intensity within and outside of the cropping system should be performed near midday, when the sun is as close as possible to directly overhead for the latitude and season. In addition, there should be as little cloud cover as possible and no strong winds.

Coordination with Other Investigations

This investigation can be integrated with Investigation 19. The light environments of the three types of plots grown for that investigation can be examined and compared to both fulfill the objectives of this investigation and provide potentially valuable data for analyzing the results of the intercrop study.

Materials, Equipment, and Facilities

30 m tape measures

Compass

Fishing pole or dowel fitted with a screw eye at one end

Fishing weight (e.g., for ocean fishing) or surveyor's plumb bob

Several meters of string or cord

Table of random numbers or random number generator

Light meter (photoradiometer), preferably dual sensor, with capacity for reading photosynthetically active radiation (PAR) (if a light meter is not available, see Variations and Further Study, suggestion #3)

Electronic data logger, for use with the light meter (optional)

Advance Preparation

- Collect materials.
- Identify three appropriate cropping systems for study. The systems (or parts thereof) selected can be relatively small scale, such as different beds within a diverse garden-scale system. The systems should differ in one or more of the following respects: crop type, crop age, planting density, canopy structure, and species diversity. They should also be similar in exposure and slope.

Very different kinds of systems, such as orchards, vineyards, and annual cropping systems, can be compared in this investigation. In general, the more variation, the more interesting the data.

A set of suitable agroecosystems might exist in a homegarden, an intensive organic farming operation, the fields of an agricultural experiment station, or even a community or school garden. The three types of plots planted for Investigation 19 are also suitable candidates.

Ongoing Maintenance

None required.

Investigation Teams

Form three teams, each with three to five members. Each team will be responsible for measuring the canopy and light environment of one agroecosystem. If there are more than about 18 students, a fourth group can be formed to study a fourth system. If the investigation is integrated with the intercrop overyielding investigation (#19), the four groups of that investigation may be used: one group studies one of the monocrop plots in its block, another group studies the other monocrop plot (the one in its block), and two groups study the intercrop plots in their blocks.

Procedure

Data Collection

The following steps describe the procedure for collecting canopy structure and internal light environment data in one system:

1. Determine the LAI at 20 different random points within the system. LAI is a measure of the total surface area of leaves in the 3D space above a certain 2D area of ground. The overall, or average, LAI of a cropping system gives an indication of the density of the canopy of the system, which is a major determinant of how the system as a whole utilizes sunlight.

   The method of determining LAI described in the following is called the point-intercept method. In essence, it reduces the 3D area of space within which the leaf surface area is measured to a 2D line above a 1D point on the ground. (*Note*: Instruments are available that can measure LAI directly, along with PAR; if available, such an instrument can be substituted for this manual method.)

   a. Construct an LAI measuring device. Find a pole or dowel approximately 2 m in length. Attach a screw eye near one end of the pole. (If a fishing pole is used, the terminal eyelet takes the place of the screw eye.) Pass a string about 3 m long through the eyelet at the end of the pole. To the distal end of the string attach a lead weight, such as a quarter-pound ocean-fishing sinker. Measuring from the end of the weight, place marks or tape on the string at 25 cm intervals.

b. Divide the cropping system (or some representative portion of it if it is very large) into rough quadrants. Mark the borders of the quadrants with stakes and/or string.

c. Randomly select a point within each quadrant. This can be done by tossing an object backward over the shoulder and marking where it lands or by generating two random numbers within an appropriate range to designate x and y coordinates within the quadrant.

d. Mark each of the four randomly selected locations, as well as the very center of the system (where the corners of all four quadrants meet). These five spots will be the locations from which sampling will take place.

e. Designate one person as the sampler. Position the sampler at one of the previously selected points, with the LAI measuring device in hand. *Note:* If the cropping system being studied is fairly small (e.g., a garden bed), this step, as well as steps f–h, may have to be modified so that the sampler can stand outside the system.

f. Designate another person as the pointer. The pointer determines north with a compass (magnetic north is adequate) and directs the sampler to stand facing north.

g. The sampler extends the pole an arbitrary distance in a northerly direction, holding the string so that the weight is above the top of the crop canopy. The weight should be within the boundaries of the quadrant.

h. With eyes closed (so as not to bias the exact location of drop), the sampler lowers the weight into the foliage until it just touches the ground. He or she then holds the pole and line steady.

i. Designate a third person as the reader. The reader examines the string and notes each contact between the string and a leaf, beginning with the one highest above the ground (contact with a stem does not count). On the LAI datasheet, the reader records, for each contact, the species or type of plant and the point of contact's distance above the ground to the nearest 25 cm.

j. Repeat steps g–i for the other three compass directions at that location.

k. Repeat steps f–j for the other four marked points in the cropping system, changing roles if the team is made up of more than three people.

l. When data have been collected for all 20 points in the system, add up the number of leaf contacts for each point where the weighted line was dropped, and record these totals in the right-hand column of the datasheet. Each measurement of the total number of contacts is the LAI for that point in the cropping system.

m. Sum the column of point-wise LAI measurements and divide this total by 20 to derive a mean LAI, which is the estimated LAI for the cropping system.

2. Measure the average amount of PAR (wavelength = 400–700 nm) transmitted through the canopy of the cropping system to the ground. The procedure should be completed at midday.

The ideal method for measuring transmitted light is to connect two sensors to a light meter via a switching box. One sensor is placed in a fixed position in a clearing just outside the cropping system, and the other is moved within the system to take shade readings at various locations. The light meter should be capable of measuring only PAR. With this setup, the full-sunlight reading (which can vary over short periods of time) can be compared almost simultaneously with each shade reading. The recording of light transmission data can be simplified with the use of an electronic data logger connected to the light meter. The following steps assume that dual sensors and a switching box are available; if they are not, see the note in "Alternative method" section.

a. Place a light sensor in a clearing near the cropping system, where it is exposed to full sunlight.

b. At an oblique angle to plant rows, extend a tape measure in a straight line along the soil surface 25 m through the cropping system. Light readings will be taken along this transect line at 50 cm intervals. Locate the transect near the center of the cropping system if possible, keeping in mind that a cable must be able to be stretched from every point on the transect to the fixed sensor in the clearing. If the cropping system is not large enough to accommodate a 25 m transect through it, use a shorter transect. However, if the transect is shorter than about 12 m, reduce the interval of measurement along it to 0.25 m (25 cm). The goal is to have at least 25 sampling locations.

c. At the distal end of the tape measure (0 m), place the second light sensor on the ground, making sure it is level. Connect it to the switching box. Connect the fixed sensor in the clearing to the switching box as well.

    d. Note the unit of measurement on the light meter (e.g., foot-candles, watts, joules, and micro-Einsteins) and record the unit at the top of the light transmittance datasheet. Use this unit in all subsequent measurements and calculations.

    e. Take a full-sunlight reading and record it on the datasheet. Immediately thereafter, take a shade reading and record it as well. The person who positions the sensor inside the cropping system should take care to avoid affecting the light reading with his or her presence (by shading it or by reflecting light off clothing).

    f. Move the shade-sampling sensor 50 cm further along the transect. Level the sensor and take both a full-sunlight reading and a shade reading and record them.

    g. Repeat the previous step until all the points along the transect have been sampled.

Alternative method: If two sensors are not available, take a full-sunlight reading, sample all the shade locations along the transect, and then take a second full-sunlight reading. Average the two full-sunlight readings and use this figure in the datasheet.

    h. On the datasheet, sum the column of full-sunlight readings and the column of shade readings along the transect. Divide each total by 50 to determine the mean value for full-sunlight intensity and the mean value for light intensity under the vegetation of the cropping system.

    i. Calculate the mean percentage of light transmission for the cropping system by dividing the mean under-vegetation light intensity by the mean full-sunlight intensity and multiplying by 100.

Data Analysis

1. Obtain other groups' data and use it to construct tables or graphs comparing the LAIs and mean percent transmittances of each system.

2. Map the LAI profile of the vegetation in each cropping system. An LAI profile, constructed from the data showing which species touched the string at what height, shows the vertical distribution of leaf area in a cropping system. The following steps describe how to construct an LAI profile for one system.

    a. For each 25 cm increment of height in the cropping system, tally up the number of contacts for each species. (The data in Figure 2.1, for example, show 9 squash contacts and 0 corn contacts at 25 cm.)

    b. Divide each of the totals derived in the previous step by the total number of contacts in the system. Then multiply each result by the mean LAI for the system. Each final result is a species-specific partial LAI for that height in the cropping system. All of the partial LAIs added together will equal the mean LAI for the system.

    c. Use the partial LAIs derived in the previous step to construct a bar graph showing the partial LAIs for each species at each height. If there is more than one species at a particular height, the bar for that height will be a stacked bar showing two or more partial LAIs. An example of an LAI profile graph, constructed from the data in Figure 2.1 (and lacking stacked bars), is shown in Figure 2.3.

Write-Up

Here are some suggestions for reporting on the results of the investigation:

- Describe each agroecosystem's canopy structure and internal light environment, based on its LAI, mean percent transmittance, and LAI profile.
- Discuss how each system's canopy structure and internal light environment are related.
- Discuss how observed differences in the internal light environments among the three systems may relate to differences in the makeup or structure of each system.
- Suggest ways of managing the internal light environment by altering the structure of the cropping system.

Quadrant Direction (C = center)	Contact #1 (Species, Height)	Contact #2 (Species, Height)	Contact #3 (Species, Height)	Contact #4 (Species, Height)	Contact #5 (Species, Height)	Total Number of Contacts (Point LAI)
1-N	Squash, 25					1
1-W	Corn, 200	Corn, 150	Corn, 125	Squash, 25		4
1-S	Corn, 225	Corn, 100				2
1-E	Squash, 50					1
2-N	Corn, 175	Corn, 150	Corn 75	Squash, 50	Squash, 25	5
2-W						0
2-S	Corn, 200	Corn, 100	Squash, 50	Squash, 50		4
2-E						0
3-N	Squash, 25					1
3-W	Corn, 100	Squash, 50	Squash, 50			3
3-S	Corn, 200	Corn, 175	Squash, 25			3
3-E	Corn, 175	Corn, 175	Squash, 25			3
4-N	Corn, 150	Corn, 150				2
4-W						0
4-S	Corn, 225	Squash, 25				2
4-E	Squash, 25					1
C-N	Corn, 200	Corn, 175	Corn, 175	Squash, 50	Squash, 50	5
C-W	Corn, 200	Corn, 175				2
C-S	Corn, 175	Corn, 175	Squash, 25			3
C-E	Corn, 200					1
					Total	43
					Mean LAI (total/20)	2.15

LAI Datasheet. Sampling Date: August 23, 2015. System: Corn–Squash Intercrop

Figure 2.1
Example of a Completed LAI Datasheet.

Reports can also discuss the efficiency of light capture, shading for weed management, how the light factor might interact with other factors of the crop environment (temperature, moisture, etc.), and how time might be a factor as the crop develops, matures, and is harvested.

Variations and Further Study

1. Use the LAI data from one or more cropping systems to calculate (and compare) additional measures of canopy structure. *Canopy patchiness* is the ratio of LAI variability to mean LAI (a measure of LAI variability is the standard deviation of all the individual LAI measurements in a system). *Cover* is the number of LAI sample points for which LAI is greater than 0, divided by the total number of sample points.

2. On the basis of the results of this investigation, form a hypothesis about the relationship between canopy structure and makeup and the light environment, and test the hypothesis with further sampling and measurement.

3. Percent transmission can be estimated if a light meter is not available. Lay the tape measure inside the cropping system as described in the investigation. Then, at determined intervals, determine if the transmitted light at that point is (1) full sunlight, (2) filtered or light shade, or (3) deep shade. This determination will be subjective, but it does give an idea of how the systems vary and compare. The technique works best in systems with broad, thick leaves and when the wind is not blowing. Data are expressed as percent of sample points at each of the three levels of light intensity.

		Light Transmittance Datasheet			
Sampling Date: July 23, 2015			System: Corn–Squash Intercrop		
Unit of Measurement: Foot-Candles					
Location on Transect (m)	Light Intensity	Full-Sunlight Intensity	Location on Transect (m)	Light Intensity	Full-Sunlight Intensity
0	207	9004	13	1,376	8,773
0.5	256	9009	13.5	244	8,698
1	1900	9018	14	207	8,723
1.5	1509	9035	14.5	210	9,139
2	1235	9042	15	1,890	9,201
2.5	267	9051	15.5	2,011	9,202
3	289	9066	16	1,119	9,202
3.5	789	9073	16.5	646	9,200
4	760	9088	17	900	9,198
4.5	902	9091	17.5	700	9,192
5	933	9099	18	865	9,191
5.5	821	9103	18.5	198	9,178
6	1309	9108	19	1,098	9,171
6.5	1400	9119	19.5	178	9,165
7	487	9132	20	675	9,155
7.5	207	9147	20.5	664	9,150
8	338	9152	21	677	9,149
8.5	992	9160	21.5	346	9,148
9	1002	9163	22	198	9,139
9.5	1670	9169	22.5	1,234	9,132
10	1888	9189	23	488	9,121
10.5	1867	9187	23.5	295	9,110
11	690	9193	24	1,006	9,104
11.5	488	9199	24.5	875	9,103
12	296	9200	Total	41,868	455,442
12.5	1266	9101	Mean	837.36	9,108.84
			Mean percent transmission of light (Mean light intensity under vegetation/mean full-sunlight intensity × 100)		9.2%

Figure 2.2
Example of a Completed Light Transmittance Datasheet.

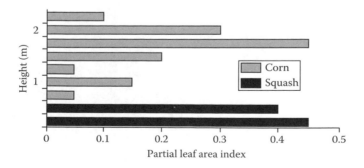

Figure 2.3
Example of an LAI Profile.

4. Set up a single crop system and take readings at weekly or 2-week intervals to see how the internal light environment and LAI change during the course of the crop cycle.

5. Measure light attenuation from the top of the crop canopy to the ground to get a more complete picture of how light varies inside the system and how this variation is determined by canopy structure. Place a fixed sensor in full sunlight as described in the investigation, but position the moveable sensor at each of several heights above each sample point on the transect. The specific heights at which readings are taken are determined by the structure of the cropping system. (See Figure 4.6 of *Agroecology: The Ecology of Sustainable Food Systems*, for an example of reading heights and attenuation data.) Once the data are used to construct a diagram such as that in the aforementioned Figure 4.6 of *Agroecology: The Ecology of Sustainable Food Systems*, each crop's role in altering the light environment can be determined.

		LAI Datasheet				

Sampling Date: System:

Quadrant Direction (C = center)	Contact #1 (Species, Height)	Contact #2 (Species, Height)	Contact #3 (Species, Height)	Contact #4 (Species, Height)	Contact #5 (Species, Height)	Total Number of Contacts (Point LAI)
1-N						
1-W						
1-S						
1-E						
2-N						
2-W						
2-S						
2-E						
3-N						
3-W						
3-S						
3-E						
4-N						
4-W						
4-S						
4-E						
C-N						
C-W						
C-S						
C-E						
					Total	
					Mean LAI (total/20)	

Light Transmittance Datasheet

Sampling Date: System:

Unit of Measurement:

Location on Transect (m)	Light Intensity	Full-Sunlight Intensity	Location on Transect (m)	Light Intensity	Full-Sunlight Intensity
0			13		
0.5			13.5		
1			14		
1.5			14.5		
2			15		
2.5			15.5		
3			16		
3.5			16.5		
4			17		
4.5			17.5		
5			18		
5.5			18.5		
6			19		
6.5			19.5		
7			20		
7.5			20.5		
8			21		
8.5			21.5		
9			22		
9.5			22.5		
10			23		
10.5			23.5		
11			24		
11.5			24.5		
12			Total		
12.5			Mean		

Mean percent transmission of light

(Mean light intensity under vegetation/mean full-sunlight intensity × 100)

Soil Temperature

Background

The temperature of the soil is an important but often overlooked variable, affecting root growth, water and nutrient uptake, root metabolism, microbial activity, decomposition of organic matter, soil chemistry, and soil moisture levels. Each type of crop plant responds differently to soil temperature, depending on its range of tolerance of soil temperatures and its temperature optimum. In agroecosystems, the goal is to maintain soil temperatures as close to the crop's optimum as possible and to keep variations in soil temperature within the crop's range of tolerance. Each combination of crop and environment, however, requires a different approach. In some situations, such as growing strawberries during winter in coastal central California, the farmer must take steps to keep the soil temperature warmer than what it would normally be. In other situations, such as growing temperate-climate vegetables in the tropics, the farmer must design and manage the system to keep the soil cool. In temperate climates, the farmer may be in the position of having to raise soil temperatures during one season and decrease them during another.

In modifying soil temperature, the farmer can take advantage of both soil coverings and the microclimate-modifying abilities of crop (and noncrop) plants themselves (Chapter 5 of *Agroecology: The Ecology of Sustainable Food Systems*). Plant leaves shade the soil, blocking absorption of solar energy and lowering the soil temperature. At night, vegetative cover has the opposite effect, helping to block convective loss of heat to the atmosphere. Soil coverings (i.e., mulches) have a more complex relationship to soil temperature. Because they are in contact with the soil, it makes a difference whether they absorb, reflect, or trap solar energy. Lighter-colored mulches will reflect much of the light that strikes them and therefore keep the soil cooler. Darker-colored mulches will absorb more solar energy as heat and transmit it to the soil below. Clear plastic (and, to some extent, materials such as floating row covers) will trap solar radiation, also raising soil temperature. At night, most mulches will reduce heat loss.

In practice, the farmer must be aware that techniques for modifying soil temperature will usually affect soil moisture as well. Mulches of all kinds, as well as increased vegetative cover, will tend to conserve soil moisture; exposing more of the soil to the sun (e.g., by pruning perennials) will dry out the soil, as well as raise its temperature.

The farmer can make inferences about soil temperature, but unlike air temperature, soil temperature is not experienced by just walking out into the field. To understand this important variable and how it is affected by agroecosystem structure and design, we must go out and measure it.

Textbook Correlation

Investigation 5: Temperature

Synopsis

The soil temperature at three depths is measured in three agroecosystems that differ in crop type, canopy structure, planting density, and/or soil cover. Temperature differences among the three systems are related to the other factors that vary among the systems.

Objectives

- Investigate the temperature of the soil and its range of variance in several agroecosystems at varying depths.
- Compare the effects of different cropping structures and/or mulches on soil temperature.
- Learn techniques for measuring soil temperature.
- Consider the effects of soil temperature on crop development.

Procedure Summary and Timeline

Data collection for this investigation can normally be completed in a single lab period.

Timing Factors

This investigation can be performed at almost any stage in the development of the cropping systems being examined.

Coordination with Other Investigations

An abbreviated version of this investigation (only 10 sample locations per system) is carried out as part of Investigation 7.

Materials, Equipment, and Facilities

Telethermometers with probes (at least three)

Metric tape measure

Three existing agroecosystems

Table of random numbers or random number generator (optional)

Advance Preparation

Identify three appropriate systems or subsystems for study. They can be relatively small scale, such as different beds within a diverse garden-scale system. These systems should

1. Differ in crop type, crop age, planting density, canopy structure, and soil surface cover (i.e., presence of mulch or presence of cover crop)
2. Be similar in exposure, slope, and soil type
3. Be located in proximity to each other

A set of suitable agroecosystems might exist in a homegarden, an intensive organic farming operation, the fields of an agricultural experiment station, or even a community or school garden. If necessary, an existing system can be made into three suitable subsystems by covering two parts of the system with different mulches a few days before temperature measurements are made.

Ongoing Maintenance

None required.

Investigation Teams

Form three teams, each with two to six members. Each team will be responsible for measuring the soil temperature of one agroecosystem or subsystem. If there are more than about 18 students, a fourth group can be formed to study a fourth system.

Procedure

Data Collection

Choose a time in the middle of a sunny day for measuring soil temperatures, if possible. The measurements in the three systems should be done simultaneously. It is interesting and instructive to take two sets of measurements about 2 h apart to see how soil temperature varies with time—see "Variations and Further Study" section.

The following steps describe the procedure for sampling one system:

1. Randomly select 40 locations within the system at which to measure soil temperature. The method for choosing the locations can vary depending on the size of the system, the time available, and the desired degree of randomness. Two basic methods are outlined here:

    Method A. Extend 10 m of a meter tape and lay it down diagonally through the system. Take temperature measurements every 25 cm along the tape.

    Method B. Establish a baseline along one edge of the plot or bed. Randomly select five points along the baseline (i.e., by generating five random numbers in the 0–100 range, each of which corresponds to the number of decimeters from one end of a 10 m baseline). From each of the five points, run a transect some distance (e.g., 8 m) into the bed or plot, perpendicular to the baseline. Along each transect, take eight temperature measurements, at each of eight evenly spaced intervals (e.g., every meter along 8 m transects).

2. At each of the locations established in step 1, use a temperature probe to measure the temperature at the soil surface, 5 cm below the surface and 15 cm below the surface. In mulched systems, the soil surface is defined as the surface of the soil proper, not the surface of the mulch or the soil covering. Thus, the mulch or covering must be carefully removed or pushed aside to measure the temperature. Follow these guidelines for measuring temperatures:

    a. To take surface readings, make sure the probe is in full contact with the soil, but not buried. A good way of ensuring this is to hold the probe nearly horizontal, with the tip touching the soil.

    b. To measure below the surface, mark the correct depth of insertion on the probe with a piece of tape or a permanent marker.

	Soil Temperature Datasheet						
Date: May 3, 2015			Time: 2 pm			Weather: Sunny	
Agroecosystem: Mulched Lettuce–Broccoli Polyculture							

Sampling Location	Temperature, °C			Sampling Location	Temperature, °C		
	Soil Surface	5 cm Depth	15 cm Depth		Soil Surface	5 cm Depth	15 cm Depth
1	26.0	17.2	14.0	22	27.5	18.3	14.6
2	35.6	22.1	16.0	23	27.3	18.3	14.5
3	27.4	19.2	14.1	24	33.8	21.0	15.2
4	30.2	20.1	14.8	25	30.2	20.0	14.8
5	33.8	21.0	15.7	26	31.9	19.2	14.9
6	31.9	19.2	14.9	27	28.2	18.0	14.6
7	27.4	18.3	14.5	28	33.5	20.4	15.0
8	28.2	18.3	14.6	29	27.2	18.1	14.3
9	34.5	21.4	15.6	30	33.9	21.7	15.8
10	33.5	21.0	15.0	31	31.5	19.1	14.4
11	28.1	18.1	14.2	32	29.2	19.1	14.2
12	27.1	18.0	14.3	33	28.0	18.4	14.6
13	26.6	17.3	14.1	34	28.5	18.6	14.8
14	31.4	19.8	15.0	35	30.0	20.1	15.3
15	33.9	21.1	15.8	36	27.3	18.2	14.4
16	28.2	18.3	14.6	37	33.4	20.5	15.1
17	30.1	21.1	14.2	38	34.3	21.1	15.5
18	30.7	20.4	14.9	39	31.4	19.3	15.8
19	34.6	21.4	15.6	40	30.1	20.0	14.7
20	33.1	20.9	15.0	Total	1223.0	784.7	594.6
21	33.5	21.1	15.2	Mean	30.58	19.62	14.87
				Standard deviation	2.81	1.35	0.55

Figure 3.1
Example of a Completed Soil Temperature Datasheet.

c. Be careful not to damage the probe when measuring below the surface. If you encounter resistance, stop and try again a few centimeters away.

d. Always be sure to wait long enough for the thermometer to reach equilibrium (a steady temperature reading). Practice using the thermometer before taking actual readings to get a feeling for how long this takes.

Record the data on the soil temperature datasheet (Figure 3.1).

Note: Measurement of the temperature at 15 cm depth can be eliminated to save time, if necessary.

Data Analysis

1. Calculate the mean for each soil depth.
2. Calculate the standard deviation for each soil depth. (Standard deviation is a measure of variation around the mean for a set of observations). If the data are entered in an electronic spreadsheet, the standard deviation can be calculated automatically with the standard deviation function in the software, or a pocket calculator with a standard deviation function can be used. Otherwise, follow the procedure described in the following.

For each set of 40 data points, calculate the "sum of squares." Each observation (x_i the soil temperature at a specific location) is subtracted from the mean ($\bar{x}$) for all 40 observations, the result is squared, and then all 40 squared results are added together. This process is easily done using a handwritten chart modeled on the one that follows.

Observation x_i	Mean $\bar{x}$	Difference $x_i - \bar{x}$	Square $\left(x_i - \bar{x}\right)^2$
	Sum of squares = $\displaystyle\sum \left(x_i - \bar{x}\right)^2$		

Use this formula to calculate the standard deviation (s) for each set of data:

$$s = \sqrt{\frac{\text{Sum of squares}}{n-1}}$$

Display, post, or otherwise publish your team's data so that each team has access to the data for all three systems.

Write-Up

The following are some suggestions for reporting on the results of the investigation:

- Describe how the soil temperatures of the three agroecosystems differ.
- Discuss how observed differences in soil temperature may relate to differences in the makeup or structure of the three systems.
- Identify the system with the most ideal soil temperature regime, and explain the reasons for it.
- Discuss observed differences in temperature range at different depths.
- Propose ideas for applying the results to the design and management of similar systems.

Variations and Further Study

1. Measure soil temperature twice in 1 day to get an impression of how temperature changes diurnally. This can easily be done at both ends of a typical 2 or 3 h lab period if enough thermometers are available. Simply keep the transects in place and measure soil temperature a second time at the same locations in each system. Ideally, the first set of data will be collected in the morning and the second set in the midafternoon, but any spacing at least 2 h apart will yield temperature differences. A more accurate picture of temperature changes can be achieved by measuring soil temperature three times during the day, at 4 h intervals. In either case, the data can then be analyzed for variation in both time and space.

2. Measure changes in soil temperature every hour over a 24 h period to understand diurnal variation in soil temperature. This type of study requires either a continual human presence (for taking temperatures every hour) or devices that can record temperature changes over time. The study can be restricted to a single system, or, if enough thermometers are available, it can be expanded to a comparison of two dissimilar systems over the same 24 h period (see, e.g., Figure 5.7 of *Agroecology: The Ecology of Sustainable Food Systems*). In either case, the best way to collect data is to set up thermometers at three to six sites in a system, at two or three depths per site, and to leave them in place, recording soil temperature hourly.

3. Measure changes in soil temperature in a single system over a growing season to see how the developing crop's modification of soil temperature changes over time. Establish permanent transects, and measure soil temperature at 20–40 sites, at two or three depths per site, every 4 weeks. Ideally, the initial measurement will occur just after seeds are sown or seedlings transplanted. This study can be performed in the three different plot types of Investigation 19.

4. Design an experimental study to test a hypothesis based on the results of this study. Such a study might be especially useful in an area with temperature extremes, where it is important to find a way of designing an agroecosystem that modifies these extremes. In northern climates, for example, designs that help raise soil temperatures may eliminate the need for greenhouses.

	Soil Temperature Datasheet						
Date:			Time:			Weather:	
Agroecosystem:							

Sampling Location	Temperature, °C			Sampling Location	Temperature, °C		
	Soil Surface	5 cm Depth	15 cm Depth		Soil Surface	5 cm Depth	15 cm Depth
1				22			
2				23			
3				24			
4				25			
5				26			
6				27			
7				28			
8				29			
9				30			
10				31			
11				32			
12				33			
13				34			
14				35			
15				36			
16				37			
17				38			
18				39			
19				40			
20				Total			
21				Mean			
				Standard deviation			

Investigation 4

Soil Moisture Content

Background

Water is of crucial importance to agriculture. In many locations, precipitation is limited, and the input of moisture it provides must either be carefully stored in the soil and conserved (Chapter 6 of *Agroecology: The Ecology of Sustainable Food Systems*) or be supplemented through irrigation (Chapter 9 of *Agroecology: The Ecology of Sustainable Food Systems*). But water that can be used for irrigation—either from surface supplies or in the ground—is also limited in supply, and agriculture must compete with industry, cities, and water-dependent natural ecosystems for its share. Moreover, irrigation itself carries with it a number of potential problems, including salinization of the soil (Chapters 1 and 9 of *Agroecology: The Ecology of Sustainable Food Systems*). For all these reasons, sustainability depends on careful water management in agroecosystems. Agroecosystems are sustainable in terms of water usage if they use water efficiently, minimize losses of soil moisture through evaporation, and are adapted to the regional climate and water supply (Chapter 9 of *Agroecology: The Ecology of Sustainable Food Systems*). Knowledge of soil moisture content and how it changes over time is a key basis for the design and management of such systems.

Textbook Correlation

Investigation 6: Humidity and Rainfall (Rainfed Agroecosystems)
Investigation 9: Water in the Soil

Synopsis

Soil moisture content in three different agroecosystems is measured, and differences in soil moisture content are related to differences in agroecosystem design and management.

Objectives

- Learn techniques for measuring soil moisture content.
- Gain firsthand knowledge of how soils vary in moisture content and moisture-holding ability.
- Compare the soil moisture content of three agroecosystems.
- Relate soil moisture retention to the aspects of agroecosystem design and management.

Procedure Summary and Timeline

Prior to week 1

- Collect equipment; choose agroecosystems to sample.

Week 1

- Collect soil samples; find wet mass of each.

Week 2

- Find dry mass of samples; calculate moisture content.

After data collection

- Analyze data; report on results.

Timing Factors

Soil samples can be collected at any time of the year, as long as the soil is not frozen or waterlogged; however, the most useful data are obtained during the growing season. In irrigated systems, samples are best collected just before irrigation is applied.

Coordination with Other Investigations

This investigation can be carried out as part of Investigation 19. Soil moisture levels in the three types of plots are measured at some point in the latter developmental stages of the crops to determine if the different cropping architectures affect soil moisture retention or soil moisture use.

An abbreviated version of this investigation (only three samples per system) is carried out as part of Investigation 7.

Materials, Equipment, and Facilities

Trowels or spades

Heat-resistant marker

Metric ruler or tape

Drying oven

Scale

Sixteen soil sample cans with lids per team (48 total)

Three different agroecosystems from which to take soil samples

Advance Preparation

Select three distinct agroecosystems from which to collect soil samples. The three systems should have identical soils and inputs of moisture but varying designs and/or management strategies. The best way to control for moisture input is to choose systems near each other, with the same slope and elevation, so that natural moisture input from precipitation is likely to be the same for all three systems. It is possible to sample irrigated systems if the input of irrigation water can be (or has been) made identical prior to the soil sampling. In such systems, sampling should be done just before an episode of irrigation, since this is when soil moisture content will differ the most among systems.

Ongoing Maintenance

The soil sample cans need to be removed from the drying oven after 24 h.

Investigation Teams

Form three teams, each with three to five members. Each team will be responsible for sampling one agro-ecosystem. If there are more than 15 students, a fourth system can be studied, or one large system can be divided into two parts.

Procedure

Data Collection

The gravimetric procedure described in the following makes use of widely available equipment and is highly accurate. Other more rapid methods of measuring soil water content are available (e.g., using neutron probes), but they all have limitations. See "Variations and Further Study" section for information on their use.

The steps that follow describe the procedure for collecting soil moisture data from one agroecosystem:

1. Secure 16 soil sample cans with lids and number them sequentially from 1 to 16 with a permanent heat-resistant marker (both can and lid should be numbered).
2. Find the mass of each can and lid to the nearest 0.1 g. Write the mass of each can directly on the can's lid.
3. Randomly select eight locations within the system at which to take samples. The method for choosing the locations can vary depending on the size of the system, the time available, and the desired degree of randomness. Two basic methods are outlined here:

   Method A. Extend 8 m of a meter tape and lay it down diagonally through the system. Take samples every 1 m along the tape.

   Method B. Establish a baseline along one edge of the plot or bed, parallel to the rows of the crop. Randomly select one point along this baseline (i.e., by generating a random number in the 0–100 range and using this number to designate the number of decimeters from one end of a 10 m baseline). From this point, run a transect some distance (e.g., 8 m) into the bed or plot, perpendicular to the baseline. Along the transect, take a sample at each of eight evenly spaced intervals (e.g., every meter along an 8 m transect). If the plot is very small, two random points can be chosen on the baseline and four samples taken from each resulting transect.

4. At each sample location, take a soil sample from the 0–15 cm depth and another from the 15–30 cm depth.
   a. Clear the surface of the soil of any weeds, mulch, or dead plant matter.
   b. Using a trowel or spade, dig a small hole 15 cm deep. From the side of this hole—top to bottom—slice off a column of soil, enough to fill sample can #1. Be sure to collect the soil uniformly through the 0–15 cm zone so as not to bias the sample toward the top or bottom of the stratum.
   c. Deepen the hole to 30 cm.

d. Collect soil uniformly from the 15–30 cm depth to fill sample can #2, using the same technique as for the shallower sample. Take care to avoid contaminating the deeper sample with soil from the shallower stratum.

e. Repeat steps a–d for the seven other sample locations in sequence, continuing the pattern of using odd-numbered cans for the 0–15 depth and even-numbered cans for the 15–30 depth.

5. Bring the sample cans back to the lab. Weigh each can on the scale, recording the total mass (can plus wet soil) to the nearest 0.1 g in the appropriate column of the soil moisture datasheet. This is a good time to also record the mass of each (empty) can, which should be already written on the can's lid.

6. Remove the lid from each can, place it under the can, and put the can and lid together in a drying oven set at 105°C. Make sure each lid stays with its can. The drying process removes the gravitational water and capillary water from the soil; most of this water is available to plants. Drying does not remove the hygroscopic water and the water of hydration (see Chapter 9 of *Agroecology: The Ecology of Sustainable Food Systems*), neither of which is available to plants.

7. After 24 h of drying, remove the cans from the drying oven and replace their lids to keep out atmospheric moisture. Caution! Cans will be very hot. Handle them with protective gloves.

8. Find the total mass of each can (can plus dry soil) to the nearest 0.1 g, and record it on the datasheet (Figure 4.1).

Soil Moisture Datasheet

Sampling Date: May 23, 2015

Agroecosystem: Monocropped Conventional Lettuce

Sample Location	Can #	Sample Depth (cm)	Mass of Sample Can (g) A	Mass of Can and Wet Soil (g) B	Mass of Can and Dry Soil (g) C	Mass of Water Lost (g) $B-C$	Mass of Dry Soil (g) $C-A$	Soil Water Content, θ_m (g Water/1 g Dry Soil) $\dfrac{B-C}{C-A}$	
1	1	0–15	35.0	183.8	160.3	23.5	125.3	0.19	
	2	15–30	35.1	206.5	169.2	37.3	134.1		0.28
2	3	0–15	35.4	189.4	161.3	28.1	125.9	0.22	
	4	15–30	34.8	205.3	168.2	37.1	133.4		0.28
3	5	0–15	35.0	188.9	163.1	25.8	128.1	0.20	
	6	15–30	35.1	201.2	168.2	33.0	133.1		0.25
4	7	0–15	35.3	178.9	162.9	16.0	127.6	0.13	
	8	15–30	34.9	203.5	169	34.5	134.1		0.26
5	9	0–15	35.1	189.3	160.8	28.5	125.7	0.23	
	10	15–30	35	202.5	167.1	35.4	132.1		0.27
6	11	0–15	34.7	191.1	163.3	27.8	128.6	0.22	
	12	15–30	35.4	202.3	164.3	38	128.9		0.29
7	13	0–15	35	189.3	163.5	25.8	128.5	0.20	
	14	15–30	34.8	208.5	168	40.5	133.2		0.30
8	15	0–15	35.4	185.4	163.5	21.9	128.1	0.17	
	16	15–30	34.9	200.6	164.3	36.3	129.4		0.28
							Total	1.55	1.93
						Mean soil water content, 0–15 cm		0.19	
						Mean soil water content, 15–30 cm			0.24

Figure 4.1
Example of a Completed Soil Moisture Datasheet.

9. For each sample, calculate the mass of the water lost in the drying process and the mass of the dry soil. Record these figures on the datasheet.

10. For each sample, calculate the mass-based soil water content (θ_m) by dividing the mass of the water lost in the drying process by the mass of the dry soil.

    Note: Soil water content can be expressed in terms of a ratio of volumes (θ_v) or in terms of a ratio of masses (θ_m). We are using the mass-based expression because mass can be measured more directly. It was standard practice in the past to express both mass-based and volume-based soil moisture content as a percentage instead of a ratio; some researchers referred to the percentage of water in moist soil, while others to the "percentage" that resulted from dividing water content by the amount of dry soil.

11. Find the mean soil water content for each depth in the agroecosystem by summing each column of water content data and by dividing the result by 8.

Data Analysis

1. Make each group's data available to all investigators.
2. Compare the figures for mean soil water content among the three agroecosystems studied, and construct a graph to show the differences.

Write-Up

Here are some suggestions for reporting on the results of the investigation:

- Describe the soil moisture conditions in each agroecosystem studied. Explain how each system's design and management affects the moisture content of its soil.
- Discuss the observed differences in soil moisture content among the three systems. Offer explanations of the differences.
- Evaluate each system in terms of water use and sustainability. Is water being used efficiently? Is soil moisture being conserved? Is excess moisture draining properly? Describe additional research that would be needed to fully answer these questions.
- Propose methods of improving soil moisture management in each of the three agroecosystems.

Variations and Further Study

1. Set up an experimental study testing how mulch affects conservation of soil moisture. Plant plots that are identical in every respect (including moisture input) except for presence or absence of mulch and chart soil moisture levels in each type of plot over time.
2. Use the methods described in this investigation as the basis for a study evaluating different types of irrigation systems (e.g., drip, sprinkler, flood, and furrow). If volumes of water applied can be measured, measurement of soil moisture levels after irrigation can yield useful data on water use efficiency.
3. Study a dry-farm agroecosystem over the growing season to see how soil moisture levels change. The data from such a study can be used to develop better ways of managing soil moisture.
4. Compare soil moisture content in soils of different type or texture (e.g., clayey soil, sandy soil, and soil high in organic matter) to gain firsthand knowledge of how soils hold moisture differently.
5. Test the water permeability of different soils by placing soil samples in cylinders, adding a known amount of water, and measuring how much water exits the bottom of the cylinder in a certain amount of time.

6. Gain a more complete understanding of the relationships among soils, plants, and moisture in a system (or systems) by measuring field capacity, permanent wilting point, and the hygroscopic coefficient. Consult a soil manual for information on equipment and technique.

7. Collect rainfall data over a year or a growing season, and correlate these data with regular measurements of soil moisture in a rainfed agroecosystem.

Note: When measuring changes in soil water content over time, indirect methods of measuring soil water content can be much more expedient. These indirect methods include using an electrical resistance block or a neutron-scattering probe. These instruments are not as accurate as the gravimetric method, but they can be calibrated with gravimetric data. After calibration, they allow quick measurement of soil moisture content and do not involve removal of soil from the field.

Soil Moisture Datasheet

Sampling Date:

Agroecosystem:

Sample Location	Can #	Sample Depth (cm)	Mass of Sample Can (g) A	Mass of Can and Wet Soil (g) B	Mass of Can and Dry Soil (g) C	Mass of Water Lost (g) $B-C$	Mass of Dry Soil (g) $C-A$	Soil Water Content, θ_m (g Water/1 g Dry Soil) $\dfrac{B-C}{C-A}$	
1	1	0–15							
	2	15–30							
2	3	0–15							
	4	15–30							
3	5	0–15							
	6	15–30							
4	7	0–15							
	8	15–30							
5	9	0–15							
	10	15–30							
6	11	0–15							
	12	15–30							
7	13	0–15							
	14	15–30							
8	15	0–15							
	16	15–30							
Total									
Mean soil water content, 0–15 cm									
Mean soil water content, 15–30 cm									

Investigation **5**

Soil Properties Analysis

Background

Soil is a complex, living, changing, and dynamic component of the agroecosystem. It is subject to alteration and can either be degraded or wisely managed. A thorough understanding of the ecology of the soil ecosystem is a key part of designing and managing agroecosystems in which the long-term fertility and productive capacity of the soil is maintained or even improved. This understanding begins with knowledge of how the soil is formed in a given ecological region and includes integration of all the components that contribute to the structure and function of the entire soil ecosystem (Chapter 8 of *Agroecology: The Ecology of Sustainable Food Systems*).

A great many biological, chemical, and physical factors determine soil quality. By measuring some of these components and determining how they respond to management in an agricultural context, a foundation for assessing the health of the soil can be established. Ultimately, indicators of sustainability can be grounded in the assessment of soil conditions and how they change as a result of the choices a farmer makes in managing the agroecosystem (Chapter 23 of *Agroecology*: *The Ecology of Sustainable Food Systems*).

For a complete introduction to the nature and properties of soils, consult a soils textbook (see the recommended reading list at the end of Chapter 8 of *Agroecology*: *The Ecology of Sustainable Food Systems*).

Textbook Correlation

Investigation 8: Soil
Investigation 23: Indicators of Sustainability (Assessment of Soil Health)

Synopsis

The soils from two or three agroecosystems are characterized by measuring a number of soil properties. On the basis of these data, the soils are compared, and differences in properties are linked to differences in management history.

Objectives

- Gain experience with a range of methodologies for measuring soil properties and assessing soil quality.
- Become familiar with different soil types and varying soil properties.
- Observe the relationship between soil conditions in an agroecosystem and the soil management—decisions that can be made by a farmer in relation to those conditions.
- Use the analysis of soil conditions to assess the impact of varying management strategies on soil—quality and, hence, sustainability.

Procedure Summary and Timeline

Prior to week 1

- Locate agroecosystems from which to take soil samples.

Week 1

- Sample soils.

Weeks 2–3

- Carry out soil factor analyses.

After data collection

- Analyze data; write up results.

Timing Factors

This investigation can be completed in a relatively short period of time and can be carried out any time of the year that soil can be sampled.

Materials, Equipment, and Facilities

Two or three different agroecosystems from which to take soil samples (see "Advance Preparation" section for desired characteristics)

(Other materials and equipment are listed in the Procedure section under each type of measurement.)

Advance Preparation

- Determine which of the soil factors will be measured (each is described in detail in Procedure). The choice of factors will depend in part on available equipment and time.
- Collect the materials and equipment needed for the chosen analyses.
- Identify two or more agroecosystems from which to take soil samples. These should be very distinct agroecosystems on a similar soil type that differ either in management strategy/history or crop type/system structure. It is also possible to compare two or three agroecosystems on dissimilar soil types, but this choice makes it difficult to link soil quality to management practices.

Ongoing Maintenance

No maintenance is required.

Investigation Teams

Form a convenient number of teams, each with three to five members, and assign each team responsibility for measurement of one, two, or three factors in both (or all three) agroecosystems. This method of dividing up the investigation is preferable to each team being responsible for a different agroecosystem because it allows for much more consistent sampling and more accurate measurement. When deciding which factors each team will measure, allow for differences in the length of time each type of measurement will require.

Procedure

Each procedure for measuring a type of soil property is described separately later. These descriptions are necessarily brief; if more information is desired, refer to the methodologies in *Methods for Assessing Soil Quality*, SSSA Special Publication 49 (1996).

General Note on Sampling

The properties of soil ecosystems vary both spatially and temporally, even on a small scale. Sampling must therefore be carefully planned and take into account the local variability that might be encountered.

Each soil property entails a different kind of sampling for its measurement. The procedures that follow describe the sampling steps only generally, requiring each team to determine the exact number of samples to take, how to ensure the representativeness of the samples, and how to take each sample. The following general principles apply to all the sampling carried out in this investigation:

- Define the horizontal boundaries of system to be sampled.
- Take samples from depths with the greatest agricultural significance (this should be the 0–15 cm upper stratum unless otherwise noted).
- Use a standard sample design that insures the samples are representative of the system. Depending on the system and the factor being measured, sampling can be random, selective (based on judgment), or stratified random (divided into subsites in which samples are taken randomly).
- Take an adequate number of samples to account for local variability.
- Use proper soil collection, handling, and sampling equipment and techniques.

Measuring pH

Equipment needed:

- pH meter (or calibrated litmus paper or pH testing reagents from a standard soil test kit)
- Convective oven for drying samples

Soil acidity or pH is a measure of the hydrogen ion (H+) activity in the soil solution, in this case in water, and is specifically defined as the $-\log_{10}$ of the hydrogen ion concentration. Soil pH will rise or fall depending on the impact of a range of factors, including farming practices. If, as a result of these impacts, pH falls below or rises above certain optimum levels for biological and chemical activity (Chapter 8 of *Agroecology: The Ecology of Sustainable Food Systems*), the soil will become much less productive.

1. Choose a sampling design.
2. Collect 5–10 soil samples from each agroecosystem.

3. Dry the samples in a convective oven at 105°C for about 24 h.

4. Construct a data table in your lab notebook to record the pH readings of the samples.

5. Run each sample through a sieve with 2 mm openings.

6. For each sample, mix equal volumes of dry soil and distilled water (1:1). A good volume to use is 30 mL. Mix the soil and water thoroughly with a glass rod until a uniform paste is formed.

7. Measure the pH of each sample paste. Any one of three different methods can be used:

   a. Measure the pH directly with a combination electrode and calibrated millivolt meter (called a pH meter).

   b. Apply the paste to calibrated litmus paper and compare the resulting color to a standard reference color sheet.

   c. Use the pH measurement procedure from a standard soil test kit, in which the color of a treated soil solution is compared with a standard reference color sheet.

   Record the pH of each sample in your lab notebook data table.

8. Record the pH of each sample in your lab notebook data table.

9. Calculate the mean pH for each agroecosystem, and record these figures in the appropriate cells of the Soil Properties Datasheet.

Measuring Electrical Conductivity

Equipment needed:

- Conductivity meter
- Convective oven for drying samples

The electrical conductivity (EC) of a soil solution is related to the number of cations and anions in the solution. The cations and anions are typically both from nonnutrient salts (e.g., chloride ions) and dissolved nutrients (e.g., nitrate). It is not possible to separate out the salinity component and the nutrient component of an EC measurement by itself. However, a very low EC generally indicates a soil relatively poor in nutrients and low in salts, and a high EC generally indicates problems with salt. For the purposes of this investigation, the measure will be used to check for soil salinity.

EC is a measurement of how well a solution conducts electricity, the opposite of its resistance to electricity, which is measured in ohms. The units of EC, therefore, are the reciprocal of ohms, or ohms^{-1}. This unit is commonly written as mhos, but since normal values are usually quite small, the most practical unit for measurement of EC is millimhos, or mmhos. A conductivity meter measures the current that passes between two electrodes spaced at 1 cm distance; thus, the readout is millimhos/cm.

1. Choose a sampling design.

2. Collect and properly label five or more soil samples from each agroecosystem.

3. Dry the samples in a convective oven at 105°C for about 24 h.

4. Construct a data table in your lab notebook to record the EC data.

5. Run each sample through a sieve with 2 mm openings.

6. For each sample, mix equal volumes of dry soil and distilled water (1:1). A good volume to use is 30 mL. Mix the soil and water thoroughly with a glass rod until a uniform paste is formed.

7. Let the paste mixtures stand for 30 min, and then measure the EC of each sample paste by inserting the electrode of the conductivity meter into the sample. Record each conductivity reading in your lab notebook.

8. Calculate the mean EC for each agroecosystem, and record these figures in the appropriate cells of the Soil Properties Datasheet.

Measuring Bulk Density

Equipment needed:

- Cylindrical core sampler
- Marked and weighed soil cans
- Knife or spatula
- Top loading balance
- Oven to dry samples at 105°C.

The bulk density (BD) of a soil is defined as the ratio of the mass (M) of oven-dried soil to its bulk volume (V), which includes the volume of the particles and the voids (pore spaces) between the particles. BD is a dynamic soil property, altered by cultivation, compression by animals or machinery, weather, and loss of organic matter. It generally increases with depth in the soil profile and normally varies from 1.0 to 1.7 g cm^{-3}. The higher the number, the more compacted the soil and the more difficult it is for roots to penetrate. The most useful and simple method for measuring BD is to cut out a cylindrical core of soil of known volume and find the mass of the dried soil.

1. Choose a sampling design.
2. Weigh and label each soil can, if not done previously. Record the mass of each can on the can itself.
3. Determine the volume in cubic centimeters of the first 15 cm of the cylindrical core sampler (volume of a 15 cm cylinder = $\pi r^2 \times 15$, where r is the radius of the cylinder).
4. Collect five or more core samples from each agroecosystem:
   a. At each sampling location, press the sampler into the soil to a premarked line at 15 cm. Remove the sampler and trim off any excess soil flush with the end of the cylinder using the knife or spatula.
   b. Empty the contents of the sampler into a labeled soil can. Either close each can or make sure its lid stays with it.
5. Dry the samples:
   a. Remove the lid from each can, place it under the can, and put the can and lid together in a drying oven set at 105°C. Make sure each lid stays with its can.
   b. After 24 h of drying, remove the cans from the drying oven and replace their lids to keep out atmospheric moisture. Caution! Cans will be very hot. Handle them with protective gloves.
6. Construct a data table in your lab notebook to record the mass of each can, the mass of each can with its contents of dry soil, the difference between the two (the mass of the soil alone), and the BD of the sample.
7. Find the mass of each can, with its lid and dry soil contents, and record these figures in your data table.
8. Find the mass of each sample of dried soil by subtracting the mass of the can from the mass of the can with its soil. Record these figures in your data table.
9. Calculate the BD of each sample using the following formula:

$$BD(g\ cm^{-3}) = \frac{Weight\ of\ oven-dried\ soil\ in\ grams}{Volume\ of\ soil\ in\ core\ sampler\ in\ cm^3}$$

The volume of the core sampler, calculated in step #3 earlier, should be the same for each sample. Record the BD of each sample in your data table.

10. Calculate the mean BD for each agroecosystem and record these data in the appropriate cells of the Soil Properties Datasheet.

Measuring Nutrient Levels

Equipment needed:

- Commercial soil test kit (e.g., LaMotte) or availability of a soil analysis lab
- Convective oven for drying samples

Agriculture has long depended on soil nutrient tests to determine if a particular soil contains an adequate supply of nutrients to meet the needs of the crop. Most analyses of nutrients are done to determine fertilizer application rates. But if sampling of nutrient availability and concentrations is done over time, these measurements can serve as indicators of sustainability (Chapter 8 of *Agroecology: The Ecology of Sustainable Food Systems*).

There are a range of very specific, technical, elaborate, and usually automated chemical analyses that can be used to measure the nutrient content of soil samples. For this investigation, however, it is not expected that students will actually do such sophisticated analysis. Such detail is more suited to a soils course. Therefore, two more practical options are recommended in the procedure described as follows:

1. Choose a sampling design.
2. In each agroecosystem, collect five or more soil samples from the 0 to 15 cm soil stratum and another 5 to 10 samples from the 15 to 30 cm stratum.
3. Dry the samples in a convective oven at 105°C for about 24 h.
4. Construct a data table in your lab notebook to record the nitrogen, phosphorus, and potassium levels of each sample.
5. Determine the nitrogen, phosphorus, and potassium content of each soil sample, using one of the following methods:
   a. Send the samples off to a reputable soil analysis lab, and ask them to analyze for total nitrogen, nitrate nitrogen, available phosphorus, and exchangeable potassium. (This is what most farmers would do.) The lab will usually send the measurements back in either units of parts per million (ppm) or as percent soil content.
   b. Use an inexpensive commercial soil test kit, such as one of those offered by LaMotte, to test each sample. The readings provided by this method are more qualitative (e.g., "high," "medium") or reflect a possible range (e.g., 5%–10%) but are very useful for comparing soils from different systems.
6. Calculate a mean level for each nutrient at each depth in each agroecosystem. (If the data are qualitative, assign a number to each level [e.g., 4 for "high," 3 for "medium"] and calculate a mean on this basis.)
7. Record the mean nutrient-level data for each depth and each agroecosystem in the appropriate cells of the Soil Properties Datasheet. (If the data are qualitative, record the descriptive word, not the number you have assigned to represent it.)

Measuring Organic Matter Content

Equipment needed:

- Convective drying oven
- Muffle furnace (capable of heating to 360°C)
- Analytical balance
- 2.0 mm soil sieve

- 100 mL beakers
- Porcelain crucibles
- Tongs or large forceps

   Organic matter plays many important roles in the soil ecosystem, all of which are of importance to sustainable agriculture (Chapter 8 of *Agroecology: The Ecology of Sustainable Food Systems*). Soil organic matter (SOM) is one of the best indicators of soil quality, especially when a baseline number can be obtained for the soil being studied before it was put into agriculture and when soil can be observed over a period of time.

   Measuring SOM content with high precision and accuracy requires sophisticated equipment and involved techniques. In the simpler procedure outlined in the following, dried soil samples are heated at high temperature, resulting in the oxidation (to carbon dioxide) of most of the carbon present. The reduction in mass that results from this heating provides a rough measurement of organic matter content. It should be noted that various nonorganic substances, such as pure carbon (charcoal), carbonate minerals, and water of hydration, can also be lost during the heating process, contributing some measurement error. (See Weil 2005 or a similar soils text for more information on this procedure).

1. Choose a sampling design.
2. Collect three or more soil samples from each agroecosystem.
3. Construct a data table in your lab notebook to record the organic matter contents of the samples.
4. Measure the organic carbon content of the samples using the weight loss on ignition method.
   a.   Air-dry the samples and screen each one through a 2.0 mm soil sieve.
   b.   Place each sample in a 100 mL beaker and dry it at 105°C in a drying oven for 24 h to remove moisture.
   c.   Immediately upon removing each beaker from the drying oven, weigh out a 10 g sample and place the sample into a labeled porcelain crucible. (Prompt weighing of samples is important because the soil will begin to absorb moisture from the atmosphere when it is removed from the oven.)
   d.   Heat the samples at 360°C for 2 h. This process is referred to as "ignition."
   e.   Measure as precisely as possible the mass of each sample.
   f.   Find the difference between the mass of each sample before the ignition heating (10 g) and the mass of the same sample after this heating. This difference is approximately equal to the organic matter content of the sample.
   g.   Express the organic matter content of each sample as a percentage of the mass of the sample before it was subjected to the ignition heating.
5. Calculate the mean organic matter content for each agroecosystem, and record these figures in the appropriate cells of the Soil Properties Datasheet.

Measuring Surface Penetrability

Equipment needed:

- Pocket penetrometer
- Meter rule

   Penetration resistance (PR) is the capacity of a soil to resist penetration by a rigid object. PR is commonly used as an indicator of soil compaction, but it varies with soil moisture content. A series of penetrating objects, known as penetrometers, have been designed to measure PR in units of pressure, typically megapascals (MPa). A higher PR indicates greater compaction. The U.S. Department of Agriculture Soil Survey staff have used such devices to develop PR classes, all in an attempt to be able to relate soil condition to soil management. There is often a very close relationship between PR and root development in soils.

1. Create a selective sampling design, in which the selection of test locations takes into account the microrelief of the soil surface, planting rows, furrows, and pathways in the crop system. The goal is to get a representative sampling of the surface penetrability of the areas of each agroecosystem in which crop plant root systems are present. Keep in mind that at each sampling location, surface penetrability will be measured at six specific sites along a 50 cm transect.

2. Create a data table in your lab notebook to record penetrometer readings for each sample location (six microsites per transect, four or five transects, two systems).

3. Measure the surface penetrability along four or five short transects in each agroecosystem. The following steps describe the procedure for one transect:

   a. Mark a 50 cm (or somewhat shorter) transect at the sample location with a meter rule.

   b. Determine the relative soil moisture content along the transect and note it in your lab manual.

   c. Move the indicator sleeve on the penetrometer to zero on the penetration scale.

   d. Take a measurement at one end of the transect. Grip the handle and push the piston needle into the soil at a slow but steady rate until the indicator line is flush with the ground surface.

   e. Remove the penetrometer, read the scale, and record the measurement.

   f. Clean the piston and return the sliding indicator to its zero position. Repeat the test five times along the 50 cm transect, at approximate intervals of 10 cm.

4. Calculate the mean surface penetrability for each transect and then for each system as a whole. Record the means in the appropriate cells of the Soil Properties Datasheet.

Measuring Soil Texture

Texture is one of the most stable attributes of the soil (Chapter 8 of *Agroecology: The Ecology of Sustainable Food Systems*), being modified only slightly by farming practices that mix the different layers, such as cultivation. Textural classes indicate how easily a soil can be worked. Soils high in sand are easier to cultivate and are termed "light," whereas soils that are difficult to cultivate and high in clay are called "heavy." Soil texture also has an impact on water and nutrient retention and availability, with clayey soils being more retentive and sandy soils being more porous and leached.

Soil texture refers to the percentage by weight of sand (particles between 0.05 and 2.0 mm), silt (0.002–0.05 mm), and clay (<0.002 mm) in a soil sample. It is based on that part of a field-dried soil sample that passes through a 2 mm sieve. (If coarse material greater than 2 mm in diameter makes up more that 15% of a field sample, the soil can be classified as gravelly or stony.) The type of soil particle (sand, silt, or clay) that makes up the highest percentage of the sample is used to describe the soil texture class. When no one of the three fractions is dominant, the textural class is loam. (Refer to Figure 8.2 of *Agroecology: The Ecology of Sustainable Food Systems* for the various textural classes used in soil science.)

Of the many methods available for determining soil texture, there are two relatively simple methods students can use. One of these, a qualitative field method based on the feel of the soil material when kneaded or rubbed between the fingers, is described later. The other is a quantitative method using special hydrometers that measure the density of a suspended soil solution over time as the soil particles settle. Tables have been developed that allow for direct conversion of the hydrometer readings into size class percentages. If soil hydrometers and glass cylinders are available, this method is a fairly accurate method for determining soil texture classes. If you prefer to use this method, consult a soils textbook (such as the manual by Weil mentioned earlier) for details.

Some experience and demonstration is needed to make the "feel method" work consistently, but once learned, it is easily and rapidly applied. Determination of soil texture by the feel method depends on the soil's resistance to deform (alter its shape by stress or by twisting). If a moist soil is easily deformed and forms a weak ribbon when rubbed between the fingers, it is loam. Clay soils are harder to deform and make strong ribbons. Sand feels gritty and silt gives a smooth, flour- or talc-like feel. Soils high in organic matter or in a particular kind of clay mineral are more difficult to classify using the feel method and may require the hydrometer method for more accurate determination.

1. Develop a sampling design. The more homogenous the soils of the systems to be sampled, the fewer samples required.

2. Perform the following tests on each soil sample:

   a. Place about 25 g of soil in the palm of the hand. Add water drops while kneading the soil to break aggregates. Stop adding water when the soil is moldable.

   b. Shape the soil into a ball by squeezing. If it does not form a ball, it is predominantly sand. If it forms a ball, go to the next step.

   c. Hold the ball of soil between thumb and forefingers, pushing and squeezing upward to form the ball into a ribbon. If the soil does not form a ribbon, it is loamy sand, silt loam, or coarse silt loam (the hydrometer method may be needed for more specificity). If the soil forms a ribbon, go on to the next step.

   d. If the soil forms a ribbon 2.5 cm long before breaking, wet and rub the ribbon with the forefingers. If the soil feels gritty, it is sandy loam, and if it is very smooth with no grit at all, it is silt. If it feels smooth with only a little bit of grittiness, it is silt loam, and if neither grittiness nor smoothness predominates, it is loam.

   e. If the soil forms a ribbon 2.5–5 cm long before breaking, wet and rub it as in step (d). If the soil feels gritty, it is sandy clay loam. If there is no grit and it feels smooth, it is silty clay loam. If neither grittiness nor smoothness predominates, the soil is a clay loam.

   f. If the soil forms a ribbon longer than 5 cm before breaking, wet and rub it as in step (d). The soil is sandy clay if it feels gritty, silty clay if it feels smooth, and clay if it feels neither predominantly smooth nor gritty.

Soil Properties Datasheet			
Sampling Date: October 15, 2015			
	System 1	System 2	System 3
Sample site and management history	Monocropped lettuce, conventional management, Pajaro Valley, CA, 20 years	Monocropped lettuce, organic management, straw mulch renewed w/ each cropping cycle, Pajaro Valley, CA, 8 years	
pH	6.9	6.7	
Electrical conductivity (mmhos/cm)	0.7	0.4	
Bulk density (g/cm³)	1.8	1.4	
Nitrogen content (shallow/deep)	45 ppm/50 ppm	35 ppm/32 ppm	
Phosphorus content (shallow/deep)	25 ppm/36 ppm	35 ppm/31 ppm	
Potassium content (shallow/deep)	220 ppm/245 ppm	350 ppm/340 ppm	
Organic matter content (%)	1.1	2.4	
Surface penetrability (MPa)	4	1	
Soil texture	Sandy clay loam (soil ribbon 3.5 cm and gritty feel)	Sandy clay loam (soil ribbon 3.8 cm and gritty feel)	

Figure 5.1
Example of a Completed Soil Properties Datasheet.

3. Record the textural class of each sample, along with the characteristics that indicate that textural class, in your lab notebook.

4. Considering both the range of textural classes encountered in the soil samples and the predominant class, develop a description of the overall soil texture of each agroecosystem and record these descriptions in the appropriate cells of the Soil Properties Datasheet.

Write-Up

The following are some suggestions for reporting on the results of the investigation:

- Describe each soil as exhaustively as possible, including properties and characteristics such as geologic origin, color, and management history, as well as the properties you have measured.

- Describe any indicators of changes that have occurred in the soil as a result of management practices.

- Explain the observed condition of each soil property for each agroecosystem in terms of soil origin, management practices, and any other relevant factors of the soil ecosystem.

- Evaluate the overall condition and "health" of each soil. Are soil properties being maintained at sustainable levels?

- Explain how specific properties of each soil could be improved, either with changes in cropping practices or inputs. Keep in mind the design and management of a sustainable agroecosystem.

Variations and Further Study

1. If possible, sample soils from a location that has not been farmed or at least not farmed for an extended period of time. This location should be near the sampled farmed sites and on the same type of soil. Consider this unfarmed "treatment" as a control or baseline for the different managed systems. How similar or different are the managed soils after their history of farming? Do the differences give any indications of the sustainability of the farming systems?

2. Measure the field capacity, permanent wilting point, and hygroscopic coefficient of each soil, in order to gain further indications of the impacts of the different farming systems on the soil ecosystem. (Consult a soils textbook or manual for methodologies.)

3. Interview the farmers or farm managers of the sampled systems to gain a better understanding of their histories. Ask about crops grown, practices employed, annual inputs, type of machinery used, harvests, fallow treatments, weather patterns, and any other information that would be of importance in identifying factors that might impact soil quality over time.

4. If the facilities or capability are available, have the soils from the different systems analyzed for a range of other nutrients, especially essential trace elements or micronutrients that may be useful indicators of soil quality.

5. Conduct this investigation several years in succession on the same agroecosystems to track changes in soil quality. Correlating changes in soil quality with changes in the design and management of the systems may be especially instructive.

Reference

Weil, R. R., 2005, *Laboratory Manual for Introductory Soils*, 7th edn.; Dubuque, IA: Kendal-Hunt Publishing Co.

Soil Properties Datasheet			
Sampling Date:			
	System 1	System 2	System 3
Sample site and management history			
pH			
Electrical conductivity (mmhos/cm)			
Bulk density (g/cm³)			
Nitrogen content (shallow/deep)			
Phosphorus content (shallow/deep)			
Potassium content (shallow/deep)			
Organic matter content (%)			
Surface penetrability (MPa)			
Soil texture			

Canopy Litterfall Analysis

Background

In natural ecosystems, perennials shed some portion of their biomass each year. They naturally drop leaves, flowers, and fruit or seeds and can lose twigs and whole branches in windstorms. This shed biomass, or litterfall, can be quite significant in quantity (e.g., about 1 kg/m²/year in tropical forests) and plays an important role in recycling nutrients and biomass in natural systems, ensuring their long-term sustainability.

A critical aspect of sustainability in agroecosystems is duplicating the biomass recycling that occurs in natural systems. Although biomass can be recycled in an annual cropping system by returning crop residue to the soil (Chapter 8 of *Agroecology: The Ecology of Sustainable Food Systems*), perennials in agroecosystems offer the advantage of building recycling into the very structure of the system (Chapter 18 of *Agroecology: The Ecology of Sustainable Food Systems*). Perennials' contribution of organic matter in the form of litterfall is just one of their many benefits.

Textbook Correlation

Investigation 8: Soil (Management of Soil Organic Matter)
Investigation 18: Disturbance, Succession, and Agroecosystem Management

Synopsis

In an existing agroecosystem with a perennial component (e.g., orchard, vineyard, homegarden, and agroforestry system), litterfall catchers are set out, and the litter they catch is categorized and weighed biweekly to determine the amount of biomass that returns to the system as litter.

Objectives

- Learn a technique for measuring litterfall.
- Study in depth the canopy litterfall of a specific system.
- Investigate the contribution of perennials to biomass and nutrient cycling in agroecosystems.
- Determine a basis for estimating the ratio of recycled biomass to exported biomass.

Procedure Summary and Timeline

Prior to week 1

- Construct litterfall catchers.

Week 1

- Set out litterfall catchers.

Every 2 weeks thereafter

- Weigh and categorize collected litter.

After 8–10 weeks

- Analyze data and write up results.

Timing Factors

This investigation works best if data are collected over a long period of time, but the weekly investment of time is relatively small. Therefore, the investigation can run over the entire period of academic instruction, concurrently with several other investigations. In temperate climates and seasonally wet–dry tropical climates, the most useful data will be collected at the end of the growing season, when leaves are normally shed. (Ideally, in temperate climates, the litterfall catchers are left in place to collect material for an entire year.) In humid tropical climates, the timing is less critical, since plants normally shed leaves at a relatively constant rate.

Materials, Equipment, and Facilities

Litterfall catchers (see "Advance Preparation" section)

Scales with at least 0.1 g precision

Drying oven

An agroecosystem with a perennial component

Advance Preparation

- Select an appropriate system to perform the investigation. The system must have some permanent aboveground component. Good candidates include orchard and tree-crop systems; systems made up mainly of lower-stature perennials, such as vineyards and berry systems; mixed systems combining annuals and perennials, such as tropical homegardens; and systems that include noncrop trees or other perennials (e.g., as living fencerows, windbreaks, or ornamentals).
- Construct 8–12 litterfall catchers. A catcher is a 1 m × 1 m wooden frame with sides 10–15 cm high, a fine screen on the bottom, and short legs. The frame can be built with 2 × 8 or 2 × 6 lumber and fitted with ordinary window screening stapled to the bottom. If heavy objects (e.g., fruit) may fall into the catcher, the screen should be reinforced with wooden cross pieces placed underneath it. The legs should be approximately 0.25 m long, but may need to be shorter if the perennial canopy is low. For large systems, construct as many litter catchers as is practical, up to a maximum of about 20.

Ongoing Maintenance

No ongoing maintenance is needed beyond the biweekly collection of litter from the litter catchers.

Investigation Teams

Form any convenient number of teams, and give each team responsibility for one, two, three, or four litter catchers. Once the litter has been aggregated by litter type and species (see later), teams can divide the weighing according to either of these categories.

Procedure

Setup

Place the litter catchers in position in the system to be studied. Their locations will depend greatly on the size, structure, and composition of the system. If the perennial component is evenly distributed throughout the system (as in an orchard system), the catchers should be placed randomly. If the perennial component is patchy or intermittent (e.g., rows of trees interspersed among annuals), the placement scheme should be more deliberate to avoid placing catchers where they will collect only insignificant amounts of litter.

Data Collection

The procedure described as follows should be carried out once every 2 weeks following setup:

1. Put the contents of each litter catcher into a bag, bucket, or basket and take it to the lab or other area with a scale and a table surface.

2. Sort the collected litter from each catcher into relevant categories, such as leaves, twigs, flowers, and fruits/ seeds. If the litter is from more than one species of plant, categorize by species as well (to the extent possible). If separation by species proves difficult or too time consuming, skip this process and use only one set of rows in the datasheet.

3. Once all the material from all the catchers is categorized, combine same-category materials from all the catchers.

4. Dry all the collected material at 60°C for 48 h, keeping all the categories of material separate. (If the lab section meets only once a week, material can be removed from the oven after 48 h and stored in a dry location for weighing at the next meeting time.)

5. After drying, find the mass of each category of material and divide this figure by the number of litter catchers, so that the resulting number is a per-square-meter average for the system. Record this average on the Canopy Litterfall Datasheet. (Each column in the datasheet is for the data from a different week; write the date of sampling at the top of the column (Figure 6.1).)

Data Analysis

1. After the final sample has been dried and weighed and its data entered in the datasheet, sum each row of data in the datasheet, and find the overall total amount of litter collected per square meter. Also sum each column to determine the amount of litter collected per square meter for each sample period.

2. If appropriate, estimate the amount of litterfall per square meter per year.

3. If appropriate, calculate the total amount of litter per square meter collected from each species and the percentage of the overall litterfall per square meter that each species accounts for.

4. If appropriate, calculate the total amount of leaf litterfall per square meter, the amount of stem and twig litterfall per square meter, and so on, and find the percentage of the overall litterfall per square meter that each type of litter accounts for.

				Mean Mass (g) per Square Meter					
	Canopy Litterfall Datasheet Agroecosystem: Apple–Raspberry–Bush Bean Agroforestry System Number of Litter Catchers: 8
		Sample	1	2	3	4	5	Total
		Date of sampling	9/25	10/8	10/22	11/6	11/20	
Species 1: apple		Leaves	6.5	5.8	7.2	8.3	10.5	38.3
		Twigs/stems	1.1	0.7	2.0	1.1	1.3	6.2
		Fruit/seeds	0	0	0	0.4	0.6	1.0
		Flowers	0.3	0.2	0	0	0	0.5
Species 2: raspberry		Leaves	2.1	1.3	1.8	1.4	1.7	8.3
		Twigs/stems	0.3	0.1	0	0	0	0.4
		Fruit/seeds	0	0	0.3	0.4	0.2	0.9
		Flowers	0.3	0.2	0	0	0	0.5
Species 3: bush bean		Leaves	0.3	0.2	0.5	0.8	1.5	3.3
		Twigs/stems	0	0	0	0	0.2	0.2
		Fruit/seeds	0	0	0	0	0.3	0.3
		Flowers	0	0.2	0.3	0	0	0.5
Species 4: unidentified weeds		Leaves	0.3	0.4	0.2	0.6	1.9	3.4
		Twigs/stems	0.2	0.2	0.3	0.5	0.8	2.0
		Fruit/seeds	0	0	0	0	0	0
		Flowers	0.2	0	0	0	0	0.2
		Total	11.6	9.3	12.6	13.5	19.0	66.0

Figure 6.1
Example of a Completed Canopy Litterfall Datasheet.

Write-Up

The following are some suggestions for writing up the results of the investigation:

- Present a summary of the results, using graphs if appropriate.
- Discuss how the data represent the system studied. If the litter catchers were placed throughout the system, for example, the data represent the system as a whole; if they were placed only in areas with a perennial canopy, the data apply to those areas only.
- Discuss the significance of the canopy litterfall in the system as a source of recycled organic matter and nutrients. How does this amount of biomass compare to the amount removed from the system?
- If there is more than one perennial species in the system, evaluate the relative contribution of each to the system's litterfall.
- Apply the results of the investigation more generally to the design of sustainable agroecosystems, thinking of canopy litterfall as just one of the many benefits of a perennial component in agroecosystems.

Variations and Further Study

1. Study the canopy litterfall in more than one system, either sequentially or concurrently, and compare the amounts of litterfall in each.
2. Extend this investigation throughout an entire year to get a more accurate estimate of annual litterfall and to allow analysis of changes in litterfall rates during the year.

3. Estimate all three components of net primary productivity (NPP) in the system: new standing biomass, harvested biomass, and shed and pruned biomass remaining in the system. What percentage of NPP does each encompass?

4. Design an investigation of the role of pruned perennial biomass retained in the system as mulch and compost. In a specific system, what is the ideal way of augmenting litterfall with deliberate pruning?

5. Measure the litterfall for an entire year (as described in Variation #2) and then analyze the retained samples for nutrient content (N, P, K). Based on these data, calculate the annual nutrient input of the litterfall biomass. Compare this input with other inputs (i.e., soil amendments or fertilizers) and with harvest output to form a picture of nutrient cycling in the system.

6. Analyze litterfall samples for carbon and nitrogen content (and P and K content as well if desired), and use these data to calculate the C/N ratio of the litter input to the system. Investigate the role of this ratio in the decomposition process. How does the C/N ratio affect the soil fauna?

7. Retain samples of collected litterfall and use them the next season for empirical decomposition studies. For example, put the samples in fine-mesh nylon bags, place them in various locations in the system (e.g., on the surface, half buried, and fully buried), and then collect the bags after a period of time to measure decomposition.

	Canopy Litterfall Datasheet						
Agroecosystem:							
Number of Litter Catchers:							
		Mean Mass (g) per Square Meter					
	Sample	1	2	3	4	5	Total
	Date of sampling						
Species 1:	Leaves						
	Twigs/stems						
	Fruit/seeds						
	Flowers						
Species 2:	Leaves						
	Twigs/stems						
	Fruit/seeds						
	Flowers						
Species 3:	Leaves						
	Twigs/stems						
	Fruit/seeds						
	Flowers						
Species 4:	Leaves						
	Twigs/stems						
	Fruit/seeds						
	Flowers						
	Total						

Investigation

Mulch System Comparison

Background

Mulching is one of the most effective ways of modifying the conditions of an agroecosystem. A layer of mulch can radically change the soil temperature regime (Chapter 5 of *Agroecology: The Ecology of Sustainable Food Systems*), conserve soil moisture (Chapter 9 of *Agroecology: The Ecology of Sustainable Food Systems*), reduce weed growth (Chapter 11 of *Agroecology: The Ecology of Sustainable Food Systems*), provide habitat for beneficials, and contribute organic matter and nutrients to the soil as it decays (Chapter 8 of *Agroecology: The Ecology of Sustainable Food Systems*). Moreover, mulching can be relatively inexpensive and makes use of materials that would otherwise be considered waste.

A great variety of organic and synthetic materials can be employed as mulch. Each has its own advantages and disadvantages, and each has a different impact on the cropping environment.

Textbook Correlation

Investigation 5: Temperature (Modifying the Temperature Microclimate)
Investigation 8: Soil (Management of Soil Organic Matter)
Investigation 9: Water in the Soil (Managing Evapotranspiration)
Investigation 11: Biotic Factors (Organic Mulches Derived from Crops)

Synopsis

Three types of mulch are applied to different areas of an existing agroecosystem, and several weeks later, the soil temperatures, soil moisture levels, and weed growth in each mulched area, and in an unmulched one, are measured and compared.

Objectives

- Investigate the effects of mulches on various factors of the crop environment.
- Compare mulching materials.
- Test different waste materials for effectiveness as mulch.

Procedure Summary and Timeline

Prior to week 1

- Collect mulching materials; select a cropping system in which to apply the mulch.

Week 1

- Apply mulch.

Week 3, 4, 5, 6, or 7

- Measure soil temperature, soil moisture, and weed growth.

After data collection

- Analyze data and write up results.

Timing Factors

This investigation should be carried out during the growing season, because that is when the effects of the mulch will be most noticeable. Mulch application and data collection need not be carried out at any particular time, but the mulch is best applied in the earlier stages of crop development.

Materials, Equipment, and Facilities

An existing annual or perennial cropping system with room for four different treatment areas

Three different mulching materials, in adequate quantity (see "Advance Preparation" section)

Soil thermometers

Drying oven

Soil sample cans

Scale with 0.1 g precision

Metric rulers

Stakes and string or tape for marking treatment area boundaries

Several copies of the Mulch Treatment Weed Diversity Worksheet, per student or team

Advance Preparation

1. Locate an appropriate agroecosystem in which to conduct the investigation. A broad range of systems will work, including annual, perennial, and orchard systems and large-scale monocrops to small-scale gardens. The system should

   a. Be relatively homogenous throughout, to control for variables unrelated to the mulching

   b. Be large enough to accommodate three mulching treatments and an unmulched control, each taking up at least two rows of the system

   If appropriate existing systems are not available, it is of course possible to plant a crop in which to carry out the investigation.

2. Secure supplies of three different mulching materials. These should be inexpensive and locally available. They can be organic or synthetic. Ideally, they are considered waste materials that would otherwise be headed to the landfill. Possible mulching materials include the following:

   a. Straw

   b. Wood chips

   c. Shredded bark

   d. Sawdust

   e. Horticultural paper

    f. Newspaper

    g. Plastic sheeting

    h. Crop residues

    i. Shredded yard waste

    j. Aquatic weeds, such as water hyacinth

    k. Coffee grounds

    l. Crushed nutshells

3. Determine the amount of area to be covered by each type of mulch, and obtain the appropriate amount of material to cover the area (for loose spreadable materials, this will be approximately 1 m^3 of material for each 20 m^2 of area).

4. Make an appropriate number of photocopies of the Mulch Treatment Weed Diversity Worksheet.

Ongoing Maintenance

None required beyond that normally carried out in the system.

Investigation Teams

Form four teams, each made up of two to five students. Three teams will be responsible for applying one type of mulch and collecting data for that mulched area; one team will collect data for an unmulched area.

Procedure

Setup

In an annual system, mulch application can be performed anytime between a crop's early development (i.e., seedlings about 10 cm tall) and the middle stages of its development. In a perennial system, mulch should be applied well before crop maturity (e.g., no later than the time of flowering in an orchard system).

1. Determine and mark off the four treatment areas. Each should encompass at least two rows of crops, and all should be equal in size. Boundaries between treatments should be established midway between rows. Avoid the edges of fields, beds, and plots if possible. The maximum area of each treatment is limited only by practical considerations.

2. Apply a different mulch to each of three treatment areas, and leave one area as an unmulched control. Each mulching treatment should be uniform in thickness throughout and about 3–5 cm thick (except in the case of plastic sheeting). The treatments should be approximately equal in thickness, but if the materials vary greatly in density, it is more important that mass of mulch per unit area be equal than thickness of mulch be equal. Label each treatment area, indicating mulch material, mulch source, and date of application.

3. After the application of mulch, treat all four areas the same in terms of irrigation and pest control.

Data Collection

Data can be collected as early as 2 weeks after applying the mulch, but differences among mulches will become more apparent after 4–6 weeks. The following steps describe the data collection procedure for one treatment area (one team):

1. Measure midday soil temperature at 10 random locations in the treatment area, both at the surface (of the soil, not the mulch) and at a depth of 5 cm. Refer to the "Procedure" section in Investigation 3 for descriptions of how to randomly select locations for measurement and how to measure the temperature at each location. Record the data on a copy of the Soil Temperature Datasheet (the same one used for Investigation 3). Ignore the column for 15 cm data and the rows for sample locations 11–40.

   Note: As an alternative to selecting random locations in each treatment area, a protocol can be established for measuring the temperature at analogous locations in each treatment area—for example, 10 cm from the stems of two plants, just outside the edge of the shade cast by another two plants, at two locations in the open area between rows.

2. Measure soil moisture content at a depth of 5–15 cm at three random locations in the treatment area (measuring down from the surface of the soil, not the surface of the mulch). Refer to the Procedure section in Investigation 4 for descriptions of how to choose the sampling locations, collect the samples, take their wet mass, dry them, take their dry mass, and determine moisture content. Record the data on a copy of the Soil Moisture Datasheet (the same one used for Investigation 4). Ignore the rows for the 15–30 cm depth and for sample locations 4–8.

3. Observe and quantify the weed populations in the treatment area. Several different parameters can be measured. Choose among the following depending on weed growth, available time, and desired depth of analysis:

   a. *Number of individual weeds*: Either count all the weeds in the treatment area or estimate the total number by counting the weeds in a sample quadrat.

   b. *Total biomass of weeds*: Measure the wet or dry mass of a sample of weeds and use it to estimate the total weed biomass in the treatment area.

   c. *Weed diversity*: Determining weed diversity involves identifying (or at least distinguishing) weed species and finding the number of individuals of each species. Unless the number of weeds in each treatment area is very low, the most practical means of gathering data for calculating the index is to take several quadrat samples, determine the diversity index for each quadrat, and then find the average. Make photocopies of the Mulch Treatment Weed Diversity Worksheet, and use a different copy for recording the data from each quadrat and calculating its diversity index. The following steps outline the procedure for one of the quadrat samples (or for the whole treatment area if all the weeds are counted):

      i. Identify or distinguish each of the weed species in the quadrat and record each species' name (or provisional name) on the datasheet (Figure 7.1).

      ii. Count and record the number of individuals of each species (n).

      iii. Find the total number of individuals in the quadrat (N) and record this figure at the top of the datasheet.

      iv. For each species, calculate the figures for n/N, $\log_e n/N$, and $(n/N) \times (\log_e n/N)$.

      v. Sum all the species-wise calculations of $(n/N) \times (\log_e n/N)$, and change the sign of the result, which is the index of diversity for the sample.

4. Optional: If crop response appears to differ among the treatments, select one or more appropriate forms of plant growth/response measurement that will allow quantitative comparisons among the treatments. Possible parameters to measure include mean plant height, mean number of leaves per plant, mean plant biomass, and mass of harvested crop per unit area.

Data Analysis

1. If not completed previously, calculate the appropriate means, totals, and indices for the various parameters measured for each treatment area. Calculate standard deviations as appropriate or as directed.

2. Summarize the data for each treatment on the Mulch Comparison Summary Datasheet (Figure 7.2). Where calculated, express standard deviations as plus or minus deviations from the mean (e.g., 25°C ± 3.1°C). Leave spaces for unmeasured parameters blank and use the last row of the table to record data for any measurements not described here.

3. Construct graphs to allow comparisons among the different treatments.

Mulch Treatment Weed Diversity Worksheet				
Mulch Type: Straw				
Total Number of Individuals (N): 49			Sample: 1	
Species	**Number of Individuals (n)**	**n/N**	**$\log_e(n/N)$**	**$(n/N) \times (\log_e n/N)$**
Anagallis arvensis	6	0.12	−2.10	−0.26
Brassica campestris	3	0.06	−2.79	−0.17
Chenopodium album	7	0.14	−1.95	−0.28
Erodium cicutarium	1	0.02	−3.89	−0.08
Malva parviflora	6	0.12	−2.10	−0.26
Medicago sativa	5	0.10	−2.28	−0.23
Raphanus sativus	8	0.16	−1.81	−0.30
Avena fatua	2	0.04	−3.20	−0.13
Bromus diandrus	4	0.08	−2.51	−0.20
Unknown species A	7	0.14	−1.95	−0.28
			Diversity index = $-\Sigma(n/N)\log_e(n/N)$ =	2.18

Figure 7.1

Example of a Completed Mulch Treatment Weed Diversity Worksheet.

Mulch Comparison Summary Datasheet				
Time and Date of Data Collection: 1 pm, July 13, 2015				
Crop: Sweet Corn				
	Mulch #1: Loose Straw	**Mulch #2: Redwood Bark**	**Mulch #3: Black Plastic**	**Unmulched Control**
Mean soil temperature at surface (°C)	24 ± 1.9	20 ± 1.4	38 ± 0.8	35 ± 3.1
Mean soil temperature at 5 cm (°C)	19 ± 1.1	17 ± 0.8	29 ± 0.6	25 ± 1.2
Mean soil water content (g water/1 g dry soil)	0.16	0.18	0.20	0.12
Mean number of weeds per m²	73	19	3	192
Fresh biomass of weeds (kg/m²)	0.21	0.04	0.01	1.05
Shannon diversity of weeds	2.18	0.24	0.16	1.32
Mean plant height (m)	2.0	2.6	2.5	2.2
Mean plant biomass (kg)	1.2	1.6	1.5	1.0
Other parameters:				

Figure 7.2

Example of a Completed Mulch Comparison Summary Datasheet.

Write-Up

The following are some suggestions for reporting on the results of the investigation:

- Present graphs and/or tables summarizing the results of the investigation.
- Discuss the differences observed among the mulching treatments.
- Discuss the differences observed between each mulching treatment and the unmulched control. Propose explanations for these differences.

- Propose additional studies needed to more fully evaluate the utility, potential uses, and potential negative effects of each mulch material.
- Make recommendations for the use of mulch in sustainable agroecosystems.

Variations and Further Study

1. Investigate possible long-term effects of the different mulching treatments on the agroecosystem. Options include (1) measuring organic matter content of the soil before mulch application and at the end of the cropping season, (2) censusing the populations of beneficial soil microorganisms (e.g., earthworms), and (3) measuring microbial activity (e.g., by measuring the evolution of carbon dioxide). Study of the long-term effects of mulching could even be extended beyond one cropping season to include measures of soil erosion, accumulation of organic matter, and changes in nitrogen content.
2. Expand each mulching treatment to include three different mulch-layer thicknesses, such as 2, 4, and 6 cm.
3. Collect data more than once during the course of the cropping cycle to enable comparisons of changes over time.
4. Test different colors of plastic sheeting, from black to clear.

Soil Temperature Datasheet							
Date:			Time:		Weather:		
Agroecosystem:							

Sampling Location	Temperature, °C			Sampling Location	Temperature, °C		
	Soil Surface	5 cm Depth	15 cm Depth		Soil Surface	5 cm Depth	15 cm Depth
1				22			
2				23			
3				24			
4				25			
5				26			
6				27			
7				28			
8				29			
9				30			
10				31			
11				32			
12				33			
13				34			
14				35			
15				36			
16				37			
17				38			
18				39			
19				40			
20				Total			
21				Mean			
				Standard deviation			

Soil Moisture Datasheet

Sampling Date:

Agroecosystem:

Sample Location	Can #	Sample Depth (cm)	Mass of Sample Can (g) A	Mass of Can and Wet Soil (g) B	Mass of Can and Dry Soil (g) C	Mass of Water Lost (g) B–C	Mass of Dry Soil (g) C–A	Soil Water Content θ_m (g Water/1 g Dry Soil) $\dfrac{B-C}{C-A}$
1	1	0–15						
	2	15–30						
2	3	0–15						
	4	15–30						
3	5	0–15						
	6	15–30						
4	7	0–15						
	8	15–30						
5	9	0–15						
	10	15–30						
6	11	0–15						
	12	15–30						
7	13	0–15						
	14	15–30						
8	15	0–15						
	16	15–30						
Total								
Mean soil water content, 0–15 cm								
Mean soil water content, 15–30 cm								

Mulch Treatment Weed Diversity Worksheet				
Mulch Type:				
Total Number of Individuals (N): Sample				
Species	**Number of Individuals (n)**	**n/N**	**$\log_e(n/N)$**	**$(n/N) \times (\log_e n/N)$**
			Diversity index $= -\Sigma(n/N)\log_e(n/N) =$	

Mulch Comparison Summary Datasheet				
Time and date of data collection:				
Crop:				
	Mulch #1:	Mulch #2:	Mulch #3:	Unmulched Control
Mean soil temperature at surface (°C)				
Mean soil temperature at 5 cm (°C)				
Mean soil water content (g water/1 g dry soil)				
Mean number of weeds per m²				
Fresh biomass of weeds (kg/m²)				
Shannon diversity of weeds				
Mean plant height (m)				
Mean plant biomass (kg)				
Other parameters:				

Root System Response to Soil Type

Background

A crop plant's root structure is determined in part by its genotype, but the volume, density, surface area, and length of a plant's root system are influenced by the soil environment in which the plant grows (Chapter 8 of *Agroecology: The Ecology of Sustainable Food Systems*). Dense, compacted soil can restrict root penetration; a paucity (or a surplus) of nutrients can limit root growth. Ultimately, less-than-maximal root development can have a negative impact on crop production. By understanding how soil type affects the root response of different crop types, farmers can make better decisions about what crops to plant and how to alter soil conditions.

Textbook Correlation

Investigation 3: The Plant
Investigation 8: Soil

Synopsis

Seeds of an indicator species are sown into pots containing five different soil types. After a period of development under controlled conditions, the plants are carefully removed from the soil with the root systems intact. After the removal of soil, the volume, mass, and length of the root systems are measured and compared, and inferences are made about the effect of soil type on root development.

Objectives

- Learn a method of extracting root systems from soil.
- Learn methods of quantifying root volume and root length.
- Examine plants' root systems and study their structure.
- Compare the root systems of plants grown in different soils.
- Relate differences in root development to differences in soil type.
- Gain an understanding of some of the soil factors that affect root system development.

Procedure Summary and Timeline

Prior to week 1

- Obtain seeds and other materials.

Week 1

- Collect different soils; screen and air-dry the soil.

Week 2

- Sow seeds in pots.

Weekly

- Measure aboveground plant development (optional).

Week 6 or 7

- Remove roots from soil, wash roots, and quantify root development.

After data collection

- Analyze data and write up results.

Timing Factors

This investigation can be carried out at any time of the year—even in temperate-climate winters if a greenhouse is available and the soil outside is not frozen. Some planning and preparation is required for collecting soil and sowing the seeds of the indicator species, but because the exact age of the plants' root systems is not important when the plants are examined, the date for examining and measuring the root systems can remain flexible.

Materials, Equipment, and Facilities

Ten 1 gal pots per team (50 total)

Soil of five distinct types

Soil screens for screening soil (4 mm mesh hardware cloth mounted on a frame)

Soil screens for washing root systems (2 mm mesh)

Greenhouse or lathhouse

120–150 seeds of a chosen indicator species

Device for controlled watering

Pot labels

Rulers

Graduated cylinders

Drying oven (optional)

Scale for measuring mass to the nearest 0.1 g

Sheets of paper, size approx. 11 × 17 in., depending on root system area

Sheets of glass or clear plastic as large as the paper

Hand tally counter

Advance Preparation

- Select an indicator species and obtain seeds. The recommended species is sunflower, because it is easily obtained, develops substantial roots quickly, and is sensitive to soil type. Other annual crops that grow quickly and have substantial root systems, such as sweet corn, beans, squash, or broccoli, can also be used. Avoid root crops.

- Obtain other materials. If necessary, prepare greenhouse space and locate appropriate soil collection sites.

Note: The collection and preparation of soil (see "Procedure" section) can also be treated as advance preparation.

Ongoing Maintenance

After seeds are sown in the pots, they will need regular watering. It is important that all pots receive the same amount of water. The frequency of watering should be determined by the needs of the most porous soil (the seedlings in the pots with the soil most likely to dry out should not be allowed to undergo moisture stress).

Investigation Teams

Form five teams, with three to five members each. Each team will be responsible for one soil type. A team will set up 10 pots for its soil type (each a replicate) and will collect and analyze data from those 10 pots.

Procedure

Setup

1. Locate and collect soil samples. Five distinct soil types should be identified for collection, varying mainly in soil texture (Figure 8.2 of *Agroecology: The Ecology of Sustainable Food Systems*). Ideally, there will be loamy sand or sandy loam, silty loam or loam, clay loam, silty clay loam or silty clay, and clay. If desired, the process of identifying different soil textural types can include a simple analysis of soil texture (see Investigation 5, or a standard soils course manual). Additionally, the soils can vary in organic matter content and management history. Collect enough of each soil to fill at least ten 1 gal pots.

2. Record a description of each soil type, including textural composition, structure, and color. Also describe each soil's known management history and the physiographic characteristics of the site of its collection.

3. Screen each soil through 4 mm mesh screen to remove rocks and large organic debris. Crush large clayey clods.

4. Set out soils to air-dry (in a covered area if there is a chance of rain).

The following steps, which describe the procedure for one soil type, will normally be completed a week after steps 1–4:

5. Fill ten 1 gal pots with the team's soil type.

6. Sow two seeds in each pot, at the appropriate depth for the seed type.

7. Label each pot with team name, soil type or collection site, date, and replicate number. Place pots in a green-house or lathhouse, randomly mixing pots of different soil types on the bench.

8. Water each pot with 100 mL of water (the amount used can be more or less than 100 mL, but every pot in the investigation must receive the same amount).

The following step will normally be completed a week after steps 5–8:

9. Remove the weakest seedling in each pot. If only one seedling has emerged, leave that one.

Optional: Measurement of plants' aboveground development (height, number of leaves) can be made in the weeks prior to data collection. These data will allow examination of the relationship between root development and top development.

Data Collection

Begin this process 4–6 weeks after sowing the seeds. The following steps describe the preparation and data collection procedures for one plant.

1. Complete any ongoing measurements of plant height, leaf number, etc.

2. Cut the stem of the plant at the soil surface. Measure the fresh mass of the aboveground parts of the plant to the nearest 0.1 g. Record the mass on the Root Response Datasheet (Figure 8.1).

3. Place the aboveground parts of the plant in an oven to dry at 60°C for 48 h.

4. Carefully remove the soil–root mass from the pot. If it is hard and cemented (as might occur with heavy clay soils), place the soil–root mass into a bucket and soak for at least 15 min before moving on to step 5.

Root Response Datasheet

Date: May 18, 2016

Soil Type: Clayey Silt from Field in River Floodplain

Indicator Species: Sunflower

	Replicates										Mean	s
	1	2	3	4	5	6	7	8	9	10		
Aboveground fresh mass (g)	83.2	82.3	86.8	81.1	82.8	86.1	84	81.4	79.3	79.9	82.7	2.45
Aboveground dry mass (g)	9.0	8.9	9.4	8.8	9.0	9.4	9.1	8.8	8.6	8.7	8.9	0.27
Root fresh mass (g)	30.8	30.2	32.5	29.6	30.6	31.6	30.7	30.5	29.2	29.8	30.6	0.97
Root dry mass (g)	3.8	3.1	3.8	3.2	3.4	3.2	3.5	3.2	3.6	3.0	3.4	0.28
Root volume (cm³)	7.8	7.2	8.2	7.0	7.5	7.5	7.2	7.7	7.4	7.2	7.5	0.33
Root length (cm)	1520	1488	1591	1451	1504	1549	1516	1501	1432	1471	1502	46.26
Shoot/root ratio	2.37	2.90	2.51	2.73	2.65	2.89	2.57	2.73	2.42	2.91	2.67	0.20

Figure 8.1
Example of a Completed Root Response Datasheet.

5. Separate soil and root system.

   a. Place the root–soil mass on 2 mm mesh soil screen (a 2 mm hardware cloth mounted on a wooden frame will do).

   b. Using a high-pressure nozzle and hose, carefully wash the soil from the root system, retrieving any roots that break off.

   c. When the root system is free of soil particles, pat it dry between paper towels, repeat the process with fresh towels, and allow the root system to sit for a few minutes to dry further.

6. Measure the fresh mass of the root system to the nearest 0.1 g and record it on the datasheet.

7. Measure the volume of the root system.

   a. Fill an appropriately sized graduated cylinder about halfway with water.

   b. Record the starting volume of the cylinder to the nearest 1 mL.

   c. Place the root system into the cylinder, covering it completely with water.

   d. Determine the final volume. Subtract the starting volume from the final volume to find the displacement, which is equal to root system volume. The fluid volume in milliliters is equal to the spatial volume of the roots in cubic centimeters. Record the volume on the datasheet.

      Alternative method: Add water to a container with an overflow spout until water begins to flow out the spout. When the excess water has stopped flowing, place a graduated cylinder under the spout and immerse the root system completely. Measure the volume of water that flows into the graduated cylinder.

8. Estimate the length of the root system. To estimate this parameter, we will use a form of method first developed by Head (1966, *Journal of Horticultural Science* 41: 197–206) and modified by several others.

   a. Draw a 5 × 5 cm grid on a sheet of paper large enough to accommodate the root system spread over its surface.

   b. Dry the root system as thoroughly as possible by patting it with paper or cloth towels, repeating the process with fresh towels, and letting the root system sit for several minutes.

   c. Place the root system on the grid and arrange the roots so that they do not overlap. If necessary, cut long or many-branched roots and place them on unoccupied areas of the grid.

   d. Place a sheet of glass or plastic over the roots, gently pressing them down to hold them in position on the grid. Tape the glass or plastic sheet to the table or bench if necessary.

   e. Using a hand tally counter, systematically count the number of intersections between roots and grid lines (both horizontal and vertical).

      Note: If the total count of intersections is less than 50, perform steps a–e once again using a 2 × 2 cm grid.

   f. Use the following formula to calculate an estimated length, in centimeters, for the root system (if a 2 cm grid was used, replace the coefficient 3.93 with 1.57):

$$\text{Root length } (R) = \text{number of intersections} \times 3.93$$

Record the estimated root system length (to the nearest cm) on the datasheet.

9. After 48 h of drying, remove the aboveground parts of the plant from the oven and measure their dry mass (to the nearest 0.1 g). Record the dry mass on the datasheet.

10. Place the root system in an oven at 60°C for 48 h to dry it to a constant weight. Then take the dry mass of the root system (to the nearest 0.1 g) and record it on the datasheet.

11. Calculate the shoot/root ratio for the plant by dividing the dry mass of the aboveground part of the plant by the dry mass of the roots. (Within a single species, the poorer the soil, the lower the shoot/root ratio.)

Data Analysis

1. Calculate the mean of the replicates for each type of measurement on the datasheet.
2. Calculate the standard deviation (s) for each type of measurement.
   a. For each set of data (e.g., root volume), calculate the "sum of squares." Each observation (x_i; e.g., one plant's root volume) is subtracted from the mean ($\bar{x}$) for the set of 10 observations, the result is squared, and then all 10 squared results are added together. This process is easily done using a handwritten chart modeled on the following one:

Observation x_i	Mean $\bar{x}$	Difference $x_i - \bar{x}$	Square $\left(x_i - \bar{x}\right)^2$
	Sum of squares $= \sum \left(x_i - \bar{x}\right)^2$		

   b. Use the following formula to calculate the standard deviation (s) for each set of data.

$$S = \sqrt{\frac{\text{Sum of squares}}{n-1}}$$

3. Share data for all five soil types by posting a copy of each team's datasheet or assembling and displaying a master lab section datasheet.
4. Compare the various measures of plant and root system response across the five soil types. Graphs are recommended.
5. *Optional*: Perform statistical analyses of significance (i.e., t-tests) on the means of selected variables.

Write-Up

The following are some suggestions for reporting on the results of the investigation:

- Present a tabular or graphic summary of the data, including means and standard deviations for each soil type and species.
- Discuss any correlations between root development, plant development, and soil type.
- Discuss what each parameter of root system size (mass, volume, and length) indicates about a root system's development and ecological functioning.
- Evaluate the relative value of root data versus shoot data for making inferences about response to soil type.
- Discuss how soil components may interact to affect root development.
- Discuss how the results of the investigation can be applied to farming.

Variations and Further Study

1. Instead of collecting five different soil types, begin with one type of soil and alter it in different ways. For example, a very sandy soil can be altered with different amounts of organic amendment, such as screened compost or peat moss. Or a heavy clay soil can be mixed with sand in several different proportions (e.g., all clay, 5% sand, 10% sand, 20% sand, and 40% sand).

2. Instead of growing plants in pots, find field sites with soils that differ in texture, drainage, exposure, or management history and plant the indicator crop in these sites. Keep conditions in the various sites as uniform as possible, and after 5 or 6 weeks, dig up the plants and examine root system development as described in the investigation.

3. Instead of investigating soils that vary by texture, use soils that vary in chemical characteristics, such as pH and electrical conductivity.

4. Grow more than one indicator species in each soil type (either in pots or in the field). This allows examination of cross-species variability and provides more robust data. If there is more than one lab section, this is an ideal way to carry out this investigation.

Root Response Datasheet												
Date: Soil Type: Indicator Species:												
	Replicates										**Mean**	*s*
	1	2	3	4	5	6	7	8	9	10		
Aboveground fresh mass (g)												
Aboveground dry mass (g)												
Root fresh mass (g)												
Root dry mass (g)												
Root volume (cm^3)												
Root length (cm)												
Shoot/root ratio												

Section II

Studies of Population Dynamics in Crop Systems

Intraspecific Interactions in a Crop Population

Background

In a single-species crop population, all the individuals have the same requirements for nutrients, water, light, and other resources and have similar heights, root depths, and so on. As a result, the potential for mutually detrimental interactions (competitive interference) is very great, especially if a resource is limited. All other things being equal, the main factor affecting the level of negative interaction in a single-species population is the density of the plants. At a certain density, competitive interference among plants begins to become significant, lowering the ability of each plant to produce harvestable biomass (Chapter 14 of *Agroecology: The Ecology of Sustainable Food Systems*).

Many agronomists and farmers, concerned with maximizing the output of cropping systems, have focused on the single variable of planting density, attempting to determine the ideal density for each type of crop. In the process, they have gained important knowledge of intraspecific interactions, but virtually ignored interspecific interactions by dealing only with agricultural environments in which all other organisms are excluded. With overall agroecosystem sustainability as a broader goal, it is important to put population-level knowledge in the context of the community and the agroecosystem as a whole.

Textbook Correlation

Investigation 11: Biotic Factors
Investigation 14: The Population Ecology of Agroecosystems

Synopsis

In this version of the classic optimal-density study, radish seeds are sown at three different densities, and the growth, harvest biomass, and harvest root size of plants in each treatment are measured to determine how single-species plant density affects plant growth and development.

Objectives

- Observe the impacts of intraspecific interactions at densities above and below the established optimum.
- Investigate density-dependent plant–plant interactions in a crop population.
- Verify the optimum planting density for a specific crop plant.
- Understand the value of uniformity in agricultural crop populations.

Procedure Summary and Timeline

Prior to week 1

- Procure radish seeds and garden or field space and prepare soil and construct beds.

Week 1

- Sow seeds and water beds.

Week 2

- Survey emergence success.

Week 3

- Count leaves of selected plants.

Week 4

- Count leaves of selected plants.

Week 5

- Count leaves of selected plants.

Week 6

- Count leaves and measure harvest biomass and root size of selected plants.

After data collection

- Analyze data and report on results.

Timing Factors

This investigation must be carried out during the growing season.

Materials, Equipment, and Facilities

Approximately 20 m² of garden or field space

Approximately 7500 seeds of a short-cycle radish variety

Scale

Stakes and string for delimiting plots

4 m sticks

Four new pencils with erasers, for use as dibbles

Cardboard or thin plywood for making sowing templates

Four short metric rulers

Three copies of the Intraspecific Interactions Datasheet, per student or team

Advance Preparation

- Acquire radish seeds. Approximately 7500 seeds are needed. Select a variety that matures in 30–40 days (35 is ideal) and germinates in fewer than 7 days. The seeds should have been pretested for viability and have a germination rate of 95% or higher. If there are any doubts about the seeds' viability, test the germination rate before using them in the investigation.

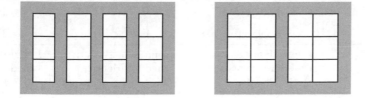

Figure 9.1
Options for layout of the beds and plots. Each plot is 1 m².

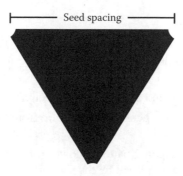

Figure 9.2
Sowing template.

- Secure appropriate garden or field space. Space is needed for constructing twelve 1 m² plots, with space between beds for easy access during sowing and sampling. The recommended layouts (see the point immediately following) will take up a space approximately 4 m wide by 4–6 m long.
- Cultivate and amend the soil as necessary, as if the radishes were to be grown for market. The soil should be fine grained, without large clods. If soil amendments are added, take care to add them uniformly over the area.
- Construct the beds and plots. It is recommended that the 12 plots be laid out in either four beds of three plots each or two beds of six plots each, as shown in Figure 9.1.

  The tops of the beds should be made as level and smooth as possible, each with no less than 1 m² of sowable space.
- Make an appropriate number of photocopies of the Intraspecific Interactions Datasheet.
- Construct three sowing templates. These are equilateral triangles made of cardboard, pressboard, or plywood, as shown in Figure 9.2. The high-density template has 3 cm sides; the medium-density template has 5 cm sides; the low-density template has 7 cm sides. The very tip of each apex (about 2 mm) should be cut off.

Ongoing Maintenance

In the first week following sowing, the seedbeds must be kept moist. Thereafter, the plots will need regular irrigation if rainfall is not adequate, as well as protection from pests (e.g., gophers, birds, and deer). If weeds germinate in the plots, they should be removed.

Investigation Teams

Form four teams, each with three to five students. Each team will be responsible for one block (set of three plots of varying density). See the following logistics map:

Team 1	Team 2	Team 3	Team 4
Block I	Block II	Block III	Block IV
1 high-density plot; 1 medium-density plot; 1 low-density plot	1 high-density plot; 1 medium-density plot; 1 low-density plot	1 high-density plot; 1 medium-density plot; 1 low-density plot	1 high-density plot; 1 medium-density plot; 1 low-density plot

Procedure

Setup

1. Decide how the three density treatments (low, medium, and high) will be distributed in the plots. The order of the treatments within each block should be both unique and random. Make a map of the chosen arrangement (Figure 9.3).

2. Label each plot with its treatment type and separate it from neighboring plots with string and stakes.

3. Make a dibble by wrapping tape around a pencil about 10 mm from the eraser end. The tape marks the depth of the holes to be made by the dibble.

4. In each plot, sow the seeds in the hexagonal pattern shown in Figure 9.4. Each seed is surrounded by six others, with the spacing between seeds determined by the size of the sowing template used (the smallest template for the high-density plot, the largest for the low-density plot, and the medium-sized template for the medium-density plot). Begin the process at one edge of a plot, using a meter stick as a straight edge for forming an initial straight line. Make holes for seeds at the apices of the template using the dibble. All the holes should be the same depth—about 10 mm. Do not cover the seeds with soil until all have been placed in holes.

5. Carefully water the beds, distributing the water as evenly as possible.

Data Collection

1. A week after sowing the seeds (week 2), check the plots for emergence of seedlings. If most of the seedlings have emerged, make a note of how many in each plot have not emerged (these will be blank spaces in the hexagonal pattern). If most seedlings have not emerged, wait until the second week to observe and make note of emergence success.

2. On the second, third, fourth, and fifth weeks after sowing (weeks 3, 4, 5, and 6), quantify the rate of growth for each plot by counting the leaves of randomly selected plants. (Note that on the final week, each plant that has its leaves counted is also harvested for another set of measurements, as described in step 3.) The following steps

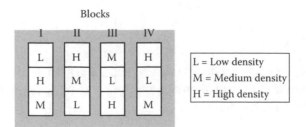

Figure 9.3
Example of a randomized distribution of the plots.

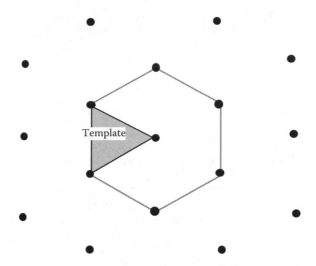

Figure 9.4
Hexagonal sowing pattern.

describe the sampling and counting procedure for one plot. A separate datasheet is used for each plot, so make two photocopies of the datasheet at the end of this investigation before you fill it in.

a.   Generate a list of 10 random numbers ranging from 1 to 99 (see the Introduction for a description of how to do this).

b.   Identify one investigator as the "counter" and one as the "observer."

c.   Beginning one row in from an edge, the counter begins counting plants out loud, moving up and down rows in some regular fashion, avoiding all plants at the edge of the plot and any plant that has fewer than six other plants surrounding it. (The counter should count every third plant in the high-density plots and every other plant in the medium-density plots.)

d.   While the counter calls out numbers, the observer compares each number with the 10 random numbers on the list. When a plant number corresponds with a random number, the observer calls out, "Do that plant!"

e.   The counter stops at each plant so identified and counts the number of true leaves on the plant 5 mm in length or longer (a short metric ruler helps in this determination). The counter calls out the number of leaves, and the observer writes the number down on a copy of the Intraspecific Interactions Datasheet (Figure 9.5). After a plant's leaves are counted, the leaf counter continues counting where he or she left off. The process continues until 10 plants have had their leaves counted.

3.   On the fifth week after sowing (week 6), harvest randomly selected plants from each plot and measure their root size and fresh biomass. The plants randomly selected for leaf counting (see step 2) on this final week are the same ones on which these measurements are made. The following steps describe the procedure for one plot:

a.   Pull up each plant selected for leaf counting, count the leaves as described in step 2, and put the whole plant in a collection bag or basket.

b.   Wash the soil off the roots of the 10 collected plants, shake off the excess water, and set the plants out to dry.

c.   When the plants are mostly dry, find the mass of each and record it in the datasheet.

d.   Measure the length of the enlarged part of the radish root (the part that is eaten) and record the length in the datasheet for the plot.

e.   Measure the width of the widest part of the radish root and record it in the datasheet for the plot.

Intraspecific Interactions Datasheet

Sowing Date: March 13, 2015

Sowing Density of Plot: Medium (5 cm Spacing)

Block: I

| Sample Plant | Number of Leaves | | | | Whole-Plant Harvest Mass (g) | Root Size | | |
	Date: 3/27	Date: 4/3	Date: 4/10	Date: 4/17		Length (mm)	Width (mm)	Index $\left(\dfrac{L \times W}{100}\right)$
1	1	2	3	4	8.6	30	18	5.4
2	1	2	4	5	11.3	27	21	5.7
3	1	2	3	4	10.6	28	20	5.6
4	2	3	4	5	15.8	32	23	7.4
5	2	3	4	5	13.4	30	23	6.9
6	2	3	4	5	15.1	28	25	7.0
7	2	3	4	5	16.0	30	26	7.8
8	1	3	4	5	11.9	25	25	6.3
9	1	2	3	4	10.8	27	23	6.2
10	1	2	4	5	11.9	27	24	6.5
Total	14	25	37	47	125.4	284	228	64.7
Mean	1.4	2.5	3.7	4.7	12.5	28.4	22.8	6.5

Number of plants in plot	835
Plot yield (kg/m²)	10.43

Figure 9.5
Example of a Completed Intraspecific Interactions Datasheet.

    f. For each sampled plant, multiply the width of the root by the length of the root and then divide by 100 to derive the root size index—a single measure of root size. Record this figure in the appropriate column of the datasheet.

    g. Count (or estimate) the number of plants in the plot and enter this number in the appropriate space at the bottom of the datasheet.

Data Analysis

1. On each datasheet, calculate the mean for each column of measurements. (Remember, these are the means for one plot of plants.)

2. If desired or directed, calculate the standard deviation for each column of measurements.

3. On each datasheet, calculate the overall yield from the plot.

    a. Multiply the average harvest biomass per plant by the number of plants in the plot.

    b. Express this result in kilograms and enter it in the appropriate space at the bottom of the datasheet. This figure is an estimate of the plot's per-square-meter yield.

4. Share the data from your team's block of plots with the other teams. Record the means for each plot on the Intraspecific Interactions Summary Worksheet (Figure 9.6). Find the overall means for each column of measurements. These means represent the averages for each density treatment across all four blocks.

5. Construct a line graph from the mean leaf-number data, showing how the average number of leaves per plant varied among the density treatments over the 4 weeks data was collected.

Intraspecific Interactions Summary Worksheet							
Low-Density Plots							
Block	**Mean Number of Leaves**				**Mean Harvest Mass (g)**	**Mean Root Size**		
	Date:	**Date:**	**Date:**	**Date:**		**Length (mm)**	**Width (mm)**	**Index**
I								
II								
III								
IV								
Total								
Overall mean								

Medium-Density Plots							
Block	**Mean Number of Leaves**				**Mean Harvest Mass (g)**	**Mean Root Size**		
	Date:	**Date:**	**Date:**	**Date:**		**Length (mm)**	**Width (mm)**	**Index**
I	1.4	2.5	3.7	4.7	12.5	28.4	22.8	6.5
II	1.0	1.9	3.3	4.5	12.1	27.0	21.9	5.9
III	1.2	2.1	3.5	4.6	12.3	27.4	23.0	6.3
IV	1.0	1.9	3.2	4.4	12.0	26.0	21.2	5.5
Total	4.6	8.4	13.7	18.2	48.9	108.8	88.9	24.2
Overall mean	1.2	2.1	3.4	4.6	12.2	27.2	22.2	6.1

High-Density Plots							
Block	**Mean Number of Leaves**				**Mean Harvest Mass (g)**	**Mean Root Size**		
	Date:	**Date:**	**Date:**	**Date:**		**Length (mm)**	**Width (mm)**	**Index**
I								
II								
III								
IV								
Total								
Overall mean								

Figure 9.6
Example of a Partially Completed Intraspecific Interactions Summary Worksheet.

6. Construct graphs showing variations in mean whole-plant harvest mass and mean root size index among the density treatments.

7. Calculate the average per-square-meter yield for each planting density among the four blocks. Compare these results with the per-plant averages for root size and harvest biomass.

Write-Up

The following are some suggestions for reporting on the results of the investigation.

- Determine to what extent plant density affected plant growth and production. Express differences in harvest mass, root size, and leaf number in a way that shows the degree of the effect.
- Compare the three types of measurements of plant growth and production as indicators of intraspecific interference.

- Compare the per-square-meter yield data with the per-plant biomass and root size data. Discuss the correlations and implications.

- Discuss variation in the data from each of the blocks.

- Discuss the new questions raised by the results and how they might be answered. Keep in mind the relationship between the design and management of the system and its sustainability.

Variations and Further Study

1. Use lettuce seedlings instead of radish seeds for the investigation. Sow the lettuce ahead of time in flats and transplant the seedlings out to the beds when they are approximately 3 cm tall. Make the plots 2 m² instead of 1 m², and plant the seedlings 15 cm apart (high density), 23 cm apart (medium density), and 31 cm apart (low density). Collect data on overall plant diameter and mass of the aboveground part of the harvested plants.

2. Alter conditions in the plots to see the effects on intraspecific interactions. For example, does doubling the amount of compost added to the soil before sowing lessen the impact of intraspecific interference? Does limiting a factor such as moisture intensify negative interactions?

3. Do not weed the plots to gain some idea of how the different planting densities affect the ability of weeds to establish.

4. Test multiline intraspecific interactions. In two of the blocks, mix equal amounts of the seed used in the other two blocks with the seed of a second variety of radish and use this mixture for sowing the plots. Compare the data from both types of blocks to determine if there is a difference in yield or performance.

5. Set up an additional set of blocks of plots, in which the identical sowing densities are used, but every other radish seed is replaced with a lettuce seedling. The radishes will quickly germinate and "catch up" to the lettuce, and the two will grow together, creating both inter- and intraspecific interactions in the plots. Take data for the radish plants as in the regular set of blocks. Comparison of the data from the radish-only and the radish–lettuce plots will shed some light on how radishes respond to being surrounded with other radishes versus being surrounded by both radishes and other species.

6. Interview local farmers to learn how they have changed or experimented with crop densities and learn how the changes have affected yield.

Intraspecific Interactions Datasheet

Sowing Date:

Sowing Density of Plot:

Block:

Sample Plant	Number of Leaves				Whole-Plant Harvest Mass (g)	Root Size		
	Date:	Date:	Date:	Date:		Length (mm)	Width (mm)	Index $\left(\dfrac{L \times W}{100} \right)$
1								
2								
3								
4								
5								
6								
7								
8								
9								
10								
Total								
Mean								

Number of plants in plot	
Plot yield (kg/m²)	

Intraspecific Interactions Summary Worksheet							
Low-Density Plots							
Block	**Mean Number of Leaves**				**Mean Harvest Mass (g)**	**Mean Root Size**		
	Date:	Date:	Date:	Date:		Length (mm)	Width (mm)	Index
I								
II								
III								
IV								
Total								
Overall mean								

Medium-Density Plots							
Block	**Mean Number of Leaves**				**Mean Harvest Mass (g)**	**Mean Root Size**		
	Date:	Date:	Date:	Date:		Length (mm)	Width (mm)	Index
I								
II								
III								
IV								
Total								
Overall mean								

High-Density Plots							
Block	**Mean Number of Leaves**				**Mean Harvest Mass (g)**	**Mean Root Size**		
	Date:	Date:	Date:	Date:		Length (mm)	Width (mm)	Index
I								
II								
III								
IV								
Total								
Overall mean								

Investigation 10

Management History and the Weed Seedbank

Background

As persistent, *r*-selected components of agroecosystems, weeds have a tendency to proliferate and compete with the crop for nutrients, water, or light. Some species can also inhibit crops allelopathically or provide habitat for herbivorous insects. To avoid serious yield losses, therefore, farmers typically devote a great deal of attention and resources to controlling weeds.

If managed properly, however, weeds have many potential benefits: they can harbor or attract beneficial arthropods, inhibit other weeds through allelopathy and competition, capture soil nutrients, control erosion, and moderate soil temperature (Chapter 16 of *Agroecology: The Ecology of Sustainable Food Systems*). These ecological roles of weeds suggest that complete elimination of weeds from agroecosystems may not be a desirable goal—even if it were a realistic one.

A farmer who seeks to manage weed populations, rather than control (or eliminate) them, must understand and work with the weeds' reproductive strategies. Weeds produce enormous numbers of seeds, which can then remain dormant in the soil for many years, forming a "weed seedbank." Since most weed seeds have a half-life of 2–5 years in the soil, management practices that consistently limit year-to-year recruitment into the seedbank can be very effective in reducing the number of weeds in the agroecosystem.

The weed seedbank is both a reflection of past management practices and an indicator of weeds' potential future in a system. Therefore, censusing and monitoring changes in the weed seedbank can be crucial in developing sustainable weed management practices.

Textbook Correlation

Investigation 11: Biotic Factors
Investigation 14: The Population Ecology of Agroecosystems
Investigation 16: Species Interactions in Crop Communities (Beneficial Interferences of Weeds)

Synopsis

Soil samples are gathered from a variety of agroecosystems and the weed seeds they contain allowed to germinate in irrigated flats. Weed seedlings from each sample are identified to species, enabling comparisons of the weed seedbanks among the sampled systems. Combined with knowledge of each systems' management history, the results also provide a basis for making inferences about the effects of different management activities (e.g., cultivation, mulching, and irrigation) on weed populations.

Objectives

- Census the weed seedbank in three different agroecosystems.
- Learn common weeds in the local area and how to identify them at the seedling and young plant stages.
- Apply techniques for measuring diversity.
- Consider the effects of different management histories on the weed seedbank and suggest strategies for sustainable weed management.

Procedure Summary and Timeline

Before week 1
- Collect materials and select soil sampling sites (see "Advance Preparation" section).

Week 1
- Collect soil samples, place in flats, and label.

Week 3
- Identify seedlings and collect initial data (optional).

Week 5
- Reidentify seedlings and collect final data.

After data collection
- Analyze data and write up results.

Timing Factors

This investigation requires only that soil containing weed seeds be available for sampling. It can therefore be performed after or before the growing season. Best results are obtained in temperate and dry-season climates, where weeds typically lie dormant for a period of time. In more humid areas, the results may reflect the more dynamic nature of weed seed production and distribution.

Materials, Equipment, and Facilities

Three different agroecosystems from which to take soil samples

Eighteen flats

Containers for soil collection in the field (e.g., four 1 gal pots per flat)

Labels

Spades (at least one per team)

Pick (to break up hard or compacted soil)

Weed and weed seedling identification guides

Vermiculite or perlite (optional, see "Ongoing Maintenance" section)

Lathhouse or greenhouse

Twelve copies of the Weed Seedbank Datasheet, per student

Six copies of the Weed Seedbank Site Diversity Worksheet, per student

Advance Preparation

- Choose three different agroecosystems from which to take soil samples. These should be easily accessible and represent a variety of management histories. Examples include pasture, field that has been fallowed for 1 year, machine-cultivated orchard, monocropped field, intercropped field, field with history of intensive tillage and herbicide use, and field with history of minimal tillage and mulching. Try to choose sites that will allow straight-forward comparisons and the generation of hypotheses regarding the effects of management.

- Gather materials: flats, containers, tools, and labels.

- Prepare space in the lathhouse or greenhouse.

- Make an appropriate number of photocopies of the Weed Seedbank Datasheet and the Weed Seedbank Site Diversity Worksheet.

Ongoing Maintenance

Study flats will need an initial watering on the day of setup. Depending on the environmental conditions, they will need watering every 2–4 days thereafter. Vermiculite or perlite can be added to the sample soil during setup at a ratio of 1:1 to reduce watering needs.

Investigation Teams

Form three teams, four to six students per team. Each team will test the soil in a different agroecosystem. (If there are enough students for four teams, a fourth system can be sampled.) In each system, team members will sample two layers of soil at each of three sites to yield a total of six soil samples to be tested for weed seed germination. See the following logistics map:

Team 1	Team 2	Team 3
System A	System B	System C
3 sample sites × 2 soil depths = 6 samples to be tested	3 sample sites × 2 soil depths = 6 samples to be tested	3 sample sites × 2 soil depths = 6 samples to be tested

Procedure

Setup

The following steps detail the procedure for sampling one agroecosystem:

1. Collect soil samples from two distinct soil layers (0–5 cm and 15–20 cm) at three sites in the system.

   a. Find a soil-collection site. Clear the vegetation and plant debris off the surface of an area approximately 0.5 m × 0.5 m. Take care to not disturb the soil itself.

   b. Using a spade, mark off a square area about 4 diameter on a side. Remove 5 cm of soil from a 1 diameter wide strip at one edge of the square. Dispose of the soil.

   c. Using the excavated strip as a point of attack, begin scraping the first 5 cm of soil from the marked-off square (Figure 10.1). Put the soil into 1 gal pots. Remove enough soil to fill four pots. Label the pots "0–5 cm, replicate 1."

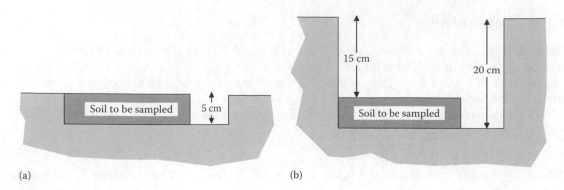

Figure 10.1
Sampling the 0–5 cm soil layer (a) and the 15–20 cm soil layer (b).

    d. Remove the remaining first 5 cm of soil from the marked-off square and dispose of it. The hole should now be a uniform 5 cm deep

    e. Remove the next 10 cm of soil from the sampling area and dispose of it. *Be very careful not to disturb the surrounding soil*; if any soil from the surface layers falls into the hole, it will contaminate the deeper layers and skew the data. A flat surface 15 cm deep should remain.

    f. Create a 20 cm deep strip along one side of the sampling area to use as a point of attack in sampling the next layer of soil.

    g. Sample the 15–20 cm soil layer as detailed in step 1c (see Figure 10.1). Label the four pots "15–20 cm, replicate 1."

    h. Repeat steps 1a–1g at two more soil-collection sites in the same system (e.g., 10 m apart in the same field). Label the pots from these locations "replicate 2" and "replicate 3."

    i. Transport the 24 pots of soil to an area suitable for transfer into flats and near the greenhouse or lathhouse.

2. When soil has been collected from all three sample sites, transfer the soil from the pots into flats.

    a. Four pots of soil from the same site and depth go into each flat.

    b. Mix the soil thoroughly in each flat and break up any large clods. (Add perlite or vermiculite at this point if desired; see "Ongoing Maintenance" section.)

    c. Label each flat with the name of the sampled system, the soil layer, and the replicate number.

3. Place the flats in the lathhouse or greenhouse, where they can be watered, receive adequate light, and be protected from environmental extremes.

4. Thoroughly water the team's six flats.

    Table 10.1 gives an overview of the flats from all three teams, as they are placed together in the lathhouse or greenhouse.

Initial Data Collection

This step is optional but recommended. It is good practice identifying weed seedlings in their early stages of growth, and it is instructive to see how weed diversity changes over time in the flats.

1. When a significant number of weed seedlings have put out their first true leaves (about 2 weeks after setup), identify the species in each study flat and count the number of individuals of each species. The following steps detail the procedure for one study flat. Two separate Weed Seedbank Datasheets will be needed for each flat

TABLE 10.1
Study Flats

Flat #	Sampled System	Soil Layer	Replicate
1	A	Surface	1
2	A	Surface	2
3	A	Surface	3
4	A	Deep	1
5	A	Deep	2
6	A	Deep	3
7	B	Surface	1
8	B	Surface	2
9	B	Surface	3
10	B	Deep	1
11	B	Deep	2
12	B	Deep	3
13	C	Surface	1
14	C	Surface	2
15	C	Surface	3
16	C	Deep	1
17	C	Deep	2
18	C	Deep	3

(one for this initial data collection and one for the final data collection), so make sure you have an appropriate number of copies of the datasheet (i.e., 12 total) before you fill it in.

a. To avoid having to sample the entire flat, mark three 10 cm diameter circular quadrats in the flat using the open end of a 1 lb coffee can. Distribute the quadrats randomly. Averages of the number of individuals per species per quadrat will serve as the basis for subsequent calculations of diversity.

b. Identify the species in each quadrat. Use whatever manuals or sources are available. Record all species names in the first column of a copy of the Weed Seedbank Datasheet. Unknown species can be named "unknown species A," "unknown species B," and so on. Record descriptions of unknown species so they can be differentiated from others and located again later. Unknown species can also be marked in the flats with small tags or toothpicks. If necessary, they can be allowed to develop until they can be identified.

c. Count the number of individuals of each species in each quadrat. Record these data on the Weed Seedbank Datasheet.

2. Calculate the average number of individuals of each species in the far-right column of the datasheet. These figures serve as measures of relative abundance of each species (even though they are averages, they are treated as discrete numbers). Sum the figures in this column to find the average total number of individuals. Figure 10.2 shows an example of a completed datasheet.

Final Data Collection

About 5 weeks after setup, collect the final data on weed diversity by following the same procedures as described for the initial data collection. To facilitate counting and identification, the seedlings in each circular quadrat can be removed from the soil and placed in same-species piles. Record the final data on fresh copies of the datasheet.

Weed Seedbank Datasheet					
Sampling Date: October 23, 2015			Agroecosystem: Pasture		
Soil Origin Depth: 0–5 cm			Replicate: 1		
Species	Number of Individuals in Quadrat A	Number of Individuals in Quadrat B	Number of Individuals in Quadrat C	Total	Average (Total/3)
Anagallis arvensis	1	0	0	1	0.33
Brassica campestris	1	1	1	3	1
Chenopodium album	1	1	1	3	1
Cirsium vulgare	2	1	3	6	2
Erodium cicutarium	0	2	1	3	1
Malva parviflora	0	0	1	1	0.33
Medicago sativa	0	0	1	1	0.33
Raphanus sativus	0	1	0	1	0.33
Sonchus oleraceus	0	1	0	1	0.33
Urtica urens	5	0	3	8	2.67
Avena fatua	1	1	1	3	1
Bromus diandrus	1	1	1	3	1
Unknown species A	2	1	3	6	2
Total average number of individuals (N)					13.33

Figure 10.2
Example of a Completed Weed Seedbank Datasheet.

Data Analysis

The data collected for each sample of soil provide a picture of the weed seedbank at the site from which the soil was sampled. To compare each weed seedbank as a community—a grouping of potential weed populations—we can determine the weed seedling diversity of each flat and compare these measures within and between sampled agroecosystems.

1. Calculate the Shannon diversity of the weed community of each study flat using the final data collected for that flat. (For a more detailed description of the Shannon index of diversity, see Chapter 17 of *Agroecology: The Ecology of Sustainable Food Systems.*)

   a. Use one copy of the Weed Seedbank Site Diversity Worksheet for each study flat. You will need six copies of the worksheet, one for each flat, so make sure you have at least five photocopies of the worksheet before you fill it in. An example of a completed worksheet is shown in Figure 10.3. At the top of the worksheet, fill in information about the soil sample site. The figure for N, the total average number of individuals for the flat, should be transferred from the flat's datasheet.

   b. For each species found in the flat, record the species name and the average number of individuals of that species (n, from the far-right column of the datasheet). Then calculate the figures for n/N, $\log_e n/N$, and $(n/N) \times (\log_e n/N)$.

   c. Sum all the species-wise calculations of $(n/N) \times (\log_e n/N)$ and change the sign of the result, to determine the index of diversity for the study flat.

Weed Seedbank Site Diversity Worksheet				
Agroecosystem: Pasture			Soil Origin Depth: 0–5 cm	
Total Average Number of Individuals (N): 13.33			Replicate: 1	
Species	Average Number of Individuals (n)	n/N	$\log_e(n/N)$	$(n/N) \times (\log_e n/N)$
Anagallis arvensis	0.33	0.03	−3.69	−0.09
Brassica campestris	1	0.08	−2.59	−0.19
Chenopodium album	1	0.08	−2.59	−0.19
Cirsium vulgare	2	0.15	−1.90	−0.28
Erodium cicutarium	1	0.03	−2.59	−0.19
Malva parviflora	0.33	0.03	−3.69	−0.09
Medicago sativa	0.33	0.03	−3.69	−0.09
Raphanus sativus	0.33	0.03	−3.69	−0.09
Sonchus oleraceus	0.33	0.03	−3.69	−0.09
Urtica urens	2.67	0.20	−1.61	−0.32
Avena fatua	1	0.03	−2.59	−0.19
Bromus diandrus	1	0.03	−2.59	−0.19
Unknown species A	2	0.15	−1.90	−0.28
Diversity index = $-\Sigma(n/N)\log_e(n/N)$ =				2.32

Figure 10.3
Example of a Completed Weed Seedbank Site Diversity Worksheet.

2. Repeat steps 1a–1c for each of the remaining study flats.

3. Calculate an estimate for the weed seedbank diversity of the sampled agroecosystem as a whole by finding the mean diversity of the three replicates at each depth. Each team member can fill in the team's data and calculate the means on a separate copy of the Weed Seedbank System Diversity Worksheet, and then the data from all three teams can be transferred to another copy of the worksheet for posting and/or photocopying and distribution (Figure 10.4).

Weed Seedbank System Diversity Worksheet						
		Site Diversity				System Diversity
Site	Depth (cm)	Replicate 1	Replicate 2	Replicate 3	Total	(Mean)
System A: Pasture	0–5	2.32	2.41	1.97	6.70	2.23
	15–20	2.87	2.68	2.82	8.37	2.79
System B: Conventional broccoli	0–5	1.19	1.23	0.98	3.40	1.13
	15–20	0.78	0.89	0.83	2.50	0.83
System C: Dry farm	0–5	0.24	0.43	0.36	1.03	0.34
	15–20	0.29	0.48	0.38	1.15	0.38

Figure 10.4
Example of a Completed Weed Seedbank System Diversity Worksheet.

Write-Up

The following are some suggestions for writing up the results of the investigation.

- Construct a graph allowing comparison of the diversity indices between systems.
- Discuss the relationship between the diversity of each system and its known management history, including input use, cultivation, and the type of crop planted. How can information about the weed seedbank be used in modifying the management of each system?
- Discuss the implications of the results for weed management in agroecosystems.
- Determine the dominant weed in each system and discuss its relationship to the system's management history.
- Present tables or graphs showing the relative abundance of each weed species by depth and by system. The decimal number in the *n/N* column of the Site Diversity Worksheet is a handy indicator of each species' relative abundance in a site and soil layer.
- Analyze and discuss the differences between the preliminary data and the final data.
- Discuss variation within systems (among the three replicates), perhaps including statistical analysis of the variation.
- Discuss the differences between soil sample depths for each agroecosystem.

Variations and Further Study

1. Use the biomass of each weed species at the final data collection date (instead of abundance) to calculate diversity.
2. Collect soil samples from a set of treatments set up in advance and designed to test a particular hypothesis. For example, the year before, grow different cover crops in a set of plots to test the effect of the crops on the weed seedbank or test the effect of different cultivation strategies.
3. Add a second layer of treatment to the experiment by manipulating one set of flats and leaving the others as a control. An example of such a treatment is adding a layer of a mulch to the surface of the treated flats before the initial watering at setup time.
4. Sample soils after a crop in the field has been planted, irrigated, and had time to begin developing. This will test for weed seeds that still might be present in the system and could form the "second wave" of weed growth.
5. In many cultural and agroecosystem contexts, some types of "weeds" are important components of sustainable systems. Design studies that can help determine what weeds may be incorporated into local systems with positive results and what types of management would be necessary.

Weed Seedbank Datasheet					
Sampling Date:			Agroecosystem:		
Soil Origin Depth:			Replicate:		
Species	Number of Individuals in Quadrat A	Number of Individuals in Quadrat B	Number of Individuals in Quadrat C	Total	Average (Total/3)
Total average number of individuals (N)					

Weed Seedbank Site Diversity Worksheet				
Agroecosystem:			Soil Origin Depth:	
Total Average Number of Individuals (N):			Replicate:	
Species	Average Number of Individuals (n)	n/N	$\log_e(n/N)$	$(n/N) \times (\log_e n/N)$
Diversity index = $-\Sigma(n/N)\log_e(n/N)$ =				

Weed Seedbank System Diversity Worksheet					
		Site Diversity				System Diversity (Mean)
Site	Depth (cm)	Replicate 1	Replicate 2	Replicate 3	Total	System Diversity (Mean)
System A:	0–5					
	15–20					
System B:	0–5					
	15–20					
System C:	0–5					
	15–20					

Comparison of Arthropod Populations

Background

Arthropod populations are highly dynamic. Mobile and abundantly reproductive, many arthropods can quickly move into a crop system, establish a population, and soon become a pest. Conventional agriculture has come to depend greatly on the use of synthetic chemical pesticides to regulate these pest populations. But in agroecosystems where the goal is to make use of natural population-regulating mechanisms (Chapter 17 of *Agroecology: The Ecology of Sustainable Food Systems*), an understanding of the population dynamics (Chapter 14 of *Agroecology: The Ecology of Sustainable Food Systems*) of each population of arthropods is critical. For this reason, population monitoring—of both pests and beneficials—has become an essential tool for managing arthropod pests.

Textbook Correlation

Investigation 14: The Population Ecology of Agroecosystems
Investigation 16: Species Interactions in Crop Communities
Investigation 17: Agroecosystem Diversity (Colonization and Diversity)

Synopsis

Arthropods are collected in three different agroecosystems, using a variety of methods, in order to determine which species are present in each system and to estimate relative population sizes. These data are used to make inferences about the relationship between agroecosystem structure/design and the arthropod populations.

Objectives

- Gain experience with several simple arthropod sampling techniques.
- Identify and learn to recognize key arthropod groups and species.
- Estimate the sizes of arthropod populations in cropping systems.
- Compare arthropod populations in three different cropping systems.

Procedure Summary and Timeline

Prior to week 1

- Locate three different crop systems and collect materials and identification guides.

Week 1

- Do initial sampling of arthropods to identify the morphospecies present in each system, set up pitfall traps, and design sampling methods.

Week 2

- Sample arthropod populations.

After sampling

- Analyze data and write up results.

Timing Factors

This investigation must be performed during the growing season. The best sampling period is just before and during the most critical period of crop development, when excessive damage from arthropod pests would cause significant reduction in yield or quality of the crop.

Coordination with Other Investigations

This investigation can be integrated into Investigation 19, as one type of comparison of the ecological dynamics in monocultures versus polycultures. For significant differences in arthropod populations to be present among the plots, however, the plots must be large enough for each plot to serve as a distinct site for arthropod population colonization and establishment, independent of arthropod populations in surrounding plots. Ideally, the plots are also separated by a distinct border or barrier, such as cultivated ground or a ground cover made up of plants very different from the crop, so that they appear as "crop islands" to pest arthropods.

Materials, Equipment, and Facilities

Three different existing agroecosystems

Six or more standard butterfly nets, some heavy and some light

2 in. diameter soil corer

Several feet of 2 in. diameter PVC pipe for making pitfall traps

Plastic cups with lip diameter slightly larger than 2 in. (they should be able to fit inside the PVC with their lips resting on the end of the PVC)

Liquid detergent

Plastic gallon jugs

Large ziplock bags

Screwtop vials (optional)

Freezer

Marking pens

Tweezers

Dissecting needles

At least three dissecting microscopes

Insect identification guides

Large petri dishes with lids (20 cm diameter)

Advance Preparation

- Identify three different agroecosystems to study. The systems should differ in management, structure, and/or type of crop. They should all be relatively large so that the main factors influencing the arthropod populations they contain are intrinsic to the system rather than related to surrounding crops or vegetation. One of the systems can be a less-disturbed noncrop area, such as a weedy field or a natural meadow or grassland. An example of three possible systems includes conventional beans, organic cabbage, and a weedy fallow field. It is also possible to select three systems in which the same crop is grown in three very different manners (e.g., conventional monoculture, organic monoculture, and organic polyculture).

- Obtain identification guides that can be used to identify key local insects, spiders, and mites, both pests and beneficials, to at least the family level. Guides should be pictorial and not rely on technical entomological knowledge.

- If the instructor does not have an entomological background, it is a good idea to ask someone who does to help out with the identification phase of the investigation.

Ongoing Maintenance

None needed if the systems studied are part of other operations.

Investigation Teams

Form three teams, each made up of four to six students. Each team will be responsible for sampling and analyzing the arthropod populations of one agroecosystem. (If time permits, each team can analyze all three agroecosystems.)

Procedure

Each team should complete the arthropod identification procedure (described immediately below) during week 1. This procedure includes planning the sampling for week 2 and setting up pitfall traps (if pitfall trapping is part of the chosen sampling methodology). During week 2, teams will use some combination of the three sampling methods described later in this section to sample the arthropod populations in their agroecosystems.

Arthropod Identification

These steps are carried out the week before sampling is to occur:

1. Gather butterfly nets and zipper-top storage bags and visit your assigned or chosen agroecosystem.
2. Take turns practicing the sweep-net sampling method (see Procedure). Collect arthropods in zipper-top storage bags.

3. Disperse into the agroecosystem and practice the visual count method of sampling (see "Procedure" section). Gain familiarity with the locations where arthropods may be hiding, on plants and on the soil surface. Collect as many different types of arthropods as possible in zipper-top storage bags. Take notes about types of arthropods and where they are found.

4. When your team has collected several bags of arthropods, return to the lab and place the bags in a freezer.

5. After 30 min in the freezer, remove the bags and empty each one into a large petri dish.

6. Separate the arthropods into same-species piles.

7. With the aid of a dissecting scope (or hand lens) and available identification guides, identify the "morphospecies" of each arthropod—a taxonomical level that adequately distinguishes the species from others. (For many groups, identification beyond genus or even family is extremely difficult.) Ask the instructor, teaching assistant, or invited expert for assistance with identification and for verifications. If the arthropods in the dishes begin to recover, place them in the freezer again.

8. In your lab notebook, record the morphospecies of each arthropod and its distinguishing characteristics. Make drawings if time allows. Give each morphospecies a unique name, even if it is as simple as "Syrphid fly sp. #1."

9. Count the number of individuals of each species or morphospecies and make a note of these numbers in your lab notebook.

10. Using what you know about each arthropod's microhabitat, morphology, behavior, and coarse taxonomy, categorize each morphospecies into a "functional group." Functional groups include the following:

    a. Predators (spiders and some types of insects, such as carabid beetles)

    b. Parasites and parasitoids (e.g., *Trichogramma* wasps)

    c. Detritus feeders (e.g., sow bugs)

    d. Scavengers (e.g., stink bugs)

    e. Omnivores (e.g., ants)

    f. Stem-boring herbivores

    g. Leaf-chewing herbivores (e.g., grasshoppers)

    h. Sucking herbivores (e.g., aphids)

11. Based on your collecting experiences, meet as a team to discuss how to best sample your agroecosystem. Design a protocol that will allow your team to estimate the relative population size of each morphospecies in the system as accurately as possible. Read through the description of each of the three sampling methods described below and decide how to incorporate them into your design. Feel free to modify each method or combine methods (e.g., you could place a net over individual plants and then use the visual count method for arthropods not caught in the net). If your team chooses to use the pitfall-trapping method, you will need to set up the traps before you begin sampling (i.e., during the remainder of the current lab period).

Pitfall-Trap Sampling

This method is the most effective for sampling ground-dwelling and nocturnal arthropods. It is more work than the other methods, but only a few well-placed traps can provide useful data.

1. One week before sampling is to begin, put the pitfall traps in place.

   a. Decide on the number of traps to construct and where to put them.

   b. Cut one 15 cm long section of 2 in. diameter PVC pipe for every trap you wish to set up. Make sure the cuts are square.

   c. At each trap location, use a 2 in. diameter (5 cm) soil corer to make a hole 15 cm deep.

   d. Push a PVC pipe section into each hole so that the end of the pipe is just below the surface of the soil.

   e. Place a plastic cup into each pipe so that its lip rests on the edge of the PVC pipe and is flush with the surface of the soil.

Arthropod Population Datasheet			
Sampling Date: May 23, 2015		Agroecosystem: Organic Pole Bean/Squash Intercrop	
Arthropod	**Functional Group**	**Sampling Method**	**Number of Individuals**
Salticid spider	Predator	Pitfall trap (5 traps)	5 (average per trap)
Carabid beetle	Predator	Pitfall trap (5 traps)	2 (average per trap)
Earwig	Omnivore	Pitfall trap (5 traps)	10.6 (average per trap)
Ant #1	Symbiont with aphids	Pitfall trap (5 traps)	40.4 (average per trap)
Ant #2	Scavenger/predator	Pitfall trap (5 traps)	8 (average per trap)
Snail	Herbivore	Visual count (15 plants)	0.8 (average per plant)
Slug	Herbivore	Visual count (15 plants)	3.4 (average per plant)
Aphidae spp.	Herbivore	Visual count (15 plants)	425 (average per plant, beans)
Phyllotreta cruciferae (flea beetle)	Herbivore	Visual count (15 plants)	5.4 (average per plant, beans)
Geocoris sp. (big-eyed bug)	Predator	Visual count (15 plants)	2.2 (average per plant)
Plutella sp. (diamondback moth, larva)	Herbivore	Visual count (15 plants)	2.8 (average per plant, beans)
Apis mellifera (honeybee)	Pollinator	Sweep net (3 transects)	15 (average per transect)
Syrphidae sp.	Predator	Sweep net (3 transects)	4.3 (average per transect)
Imported cabbage butterfly	Herbivore	Sweep net (3 transects)	2 (average per transect)
Trichogramma sp. (wasp)	Parasitoid	Sweep net (3 transects)	7.7 (average per transect)

Figure 11.1
Example of a Completed Arthropod Population Datasheet.

2. If possible, check the pitfall traps every few days and release any organisms that have fallen into them (just remove the cup and dump out the organisms).

3. Make up a soap solution in a gallon jug by filling the jug and mixing in a few milliliters of liquid detergent.

4. When sampling is to begin, fill the bottom of each trap's cup with about a centimeter of the soap solution. The soap solution will prevent any arthropod that falls in the trap from crawling out.

5. Check the interface between the soil and each cup carefully, making sure there are no gaps.

6. Leave the soapy-water cups in place for approximately 24 h. After that period of time, remove all the cups and take them to the lab. If the traps are not to be used again, remove the PVC pipe and backfill the holes.

7. In the lab, remove all the trapped arthropods from the soapy water.

8. Separate the arthropods into same-species piles. If desired, place each species into a separate vial.

9. Identify any arthropod not encountered last week and determine what functional group it belongs in.

10. Record the name of each species or morphospecies in a data table in your lab notebook. As before, use the best unique name possible.

11. For each species or morphospecies, count the number of individuals from each trap and record this figure in your lab notebook.

12. Find the total number of individuals of each species captured in all the traps. Calculate the average number of individuals of each species per trap and record this figure in the Arthropod Population Datasheet (Figure 11.1). Make a note in the "Sampling Method" column of the datasheet that pitfall trapping was used, along with the number of traps used.

Visual Count Sampling

This method is the most effective for sampling sedentary, flightless organisms such as aphids and caterpillars. It requires an ability to recognize morphospecies in the field.

1. Devise a sampling protocol. You can sample randomly selected plants or exhaustively inspect quadrats.

2. In each quadrat or on each plant, search every possible plant surface.

3. Record every morphospecies encountered and make a tally of numbers of individuals by plant or by quadrat.

4. If you encounter an arthropod not previously identified, collect it and bring it back to the lab for closer inspection and possible identification.

5. Record the name of each species encountered with this method on the datasheet. For each species, record the population data using units appropriate to your sampling method (e.g., average number of individuals per plant or per quadrat).

Sweep-Net Sampling

This method is effective for sampling flying insects such as predaceous flies and parasitoid wasps. It has the disadvantages of being potentially damaging to the crop and awkward in certain agroecosystems.

1. Devise a set of transects along which to conduct the sweeps. Parallel lines away from the edges of the system are recommended.

2. Sample each transect as follows:

   a. Stand at one end of the transect.

   b. Sweep the net across the transect line through the upper level of the crop vegetation. When you reach the end of the sweep, lift the net out of the vegetation, twist the handle of the net to block the opening, and count "1."

   c. Immediately take a step forward and move the net back into the vegetation while simultaneously opening up the net and beginning a sweep in the opposite direction. At the end of the sweep, twist the net as before and count "2."

   d. Continue in this manner for about 24 total sweeps.

   e. After the final sweep, keep the opening of the net closed and move to a clear area.

   f. With the help of a team member, turn the net inside out and empty the captured arthropods into a large zipper-top storage bag. Be careful to keep the arthropods from escaping.

3. Return to the lab with the three bags of arthropods and place the bags in the freezer for 30 min.

4. Remove the bags from the freezer and empty them into separate petri dishes.

5. Separate the arthropods into same-species piles. If desired, place each species into a separate vial.

6. Identify any arthropod not encountered last week and determine what functional group it belongs in.

7. Record the name of each species or morphospecies in a data table in your lab notebook. As before, use the best unique name possible.

8. For each species or morphospecies, count the total number of individuals collected. Calculate the average number of individuals per transect for each species and record this figure in the datasheet.

9. Release the organisms after all the data have been recorded.

Data Analysis

1. Compare the number of pest species in your system with the number of beneficial species.

2. For each sampling method, compare the relative population sizes of each type of arthropod.

3. Estimate the absolute size of each population of arthropod in your agroecosystem.

4. Share your team's data with the other teams and gather other teams' data.

5. Compare the population data among the different agroecosystems.

6. Compare the species diversity of the different agroecosystems.

Write-Up

The following are some suggestions for reporting on the results of the investigation:

- Use graphs or tables to show the comparisons described earlier.

- Discuss the variations observed among the systems and offer explanations of the causes.

- For each major species of arthropod encountered, discuss the factors that seem to determine the size of its population and its presence or absence in a cropping system.

- Discuss the probable bias associated with each method of sampling. What types of arthropods is each method most likely to detect?

- Discuss the principal types of variation in arthropod populations (spatial, temporal) and ways your sampling did or did not account for them.

- Discuss the principal sources of error in the population estimates.

- Discuss how the time of year may have influenced the arthropod populations.

- Based on your categorization of the arthropods into functional groups, construct a hypothesized food web for each agroecosystem.

Variations and Further Study

1. Sample arthropod populations more than once (e.g., at early, middle, and late stages of development) so as to follow the changes in population dynamics during the development of the cropping systems.

2. Introduce a population of a beneficial insect that is a known predator or parasite on one of the more abundant pest populations present in a cropping system. Quantify the population size of the pest before the introduction and at several times after the introduction. In addition, the beneficial insect's population in the field can be monitored, and the level of pest damage to the crop can be measured.

3. Use several other trapping methodologies to compare how each method varies in effectiveness and type of insect captured. Possible trapping methods include yellow sticky traps (good for estimating the populations of several fly pests), pheromone traps (especially good for attracting populations of male lepidopteran pests that may be more active during the night), and light traps.

Arthropod Population Datasheet			
Sampling date: Agroecosystem:			
Arthropod	**Functional Group**	**Sampling Method**	**Number of Individuals**

Census of Soil-Surface Fauna

Background

In most terrestrial ecosystems, the soil surface—the zone from the uppermost, organic matter–rich layer of soil to the top of the litter or humus layer—supports a diverse community of organisms playing important roles in the dynamic processes of the entire system. In addition to a variety of plants, this community is made up of populations of spiders, mites, insects, myriapods, molluscs, annelids, roundworms, reptiles, amphibians, moles, small rodents, bacteria, protozoa, fungi, and other organisms, functioning variously as herbivores, predators, parasites, decomposers, and detritivores (Chapter 8 of *Agroecology: The Ecology of Sustainable Food Systems*).

In agroecosystems, these organisms are often very negatively impacted by agricultural practices and activities such as cultivation, pesticide applications (including herbicides, insecticides, acaricides, and fungicides intended for other target organisms), and irrigation. In general, the more disturbance in the agroecosystem, the greater the negative impacts on these populations.

From an agroecological perspective, maintaining an active and diverse community of soil-surface organisms may be of great benefit. Some of the organisms—spiders, for example—can be important predators of pest insects. The detritivores and decomposers play an important role in breaking down organic matter and cycling nutrients within the system. Other organisms, by filling niches and enhancing diversity, increase the potential for beneficial interactions and natural regulation of populations (Chapter 17 of *Agroecology: The Ecology of Sustainable Food Systems*).

Encouraging the presence of a more diverse community of soil-surface organisms in an agroecosystem is consistent with a variety of management strategies, including minimization of disturbance (Chapter 18 of *Agroecology: The Ecology of Sustainable Food Systems*), maximization of diversity (Chapter 17 of *Agroecology: The Ecology of Sustainable Food Systems*), elimination of synthetic chemical inputs, and maintaining areas of undisturbed ecosystem within the agricultural landscape. But any attempt to enhance soil-surface populations is well served by knowledge of which species are present, how large their populations are, and how those populations may respond to various types of management.

Textbook Correlation

Investigation 2: Agroecology and the Agroecosystem Concept

Investigation 8: Soil

Investigation 16: Species Interactions in Crop Communities

Investigation 17: Agroecosystem Diversity

Investigation 18: Disturbance, Succession, and Agroecosystem Management

Synopsis

Pitfall traps are placed in four different agroecosystems, and the organisms caught in the traps are identified and counted. The census data are used to characterize each system as a habitat for the various types of organisms encountered.

Objectives

- Become acquainted with local species of invertebrates that dwell on the soil surface, under mulch, and on the lower stems of plants in agroecosystems.
- Learn a method for sampling soil-surface invertebrates.
- Compare population sizes among the various soil-surface species.
- Determine what constitutes appropriate habitat for each of the species encountered.
- Relate agroecosystem structure and design to observed diversity and population sizes of soil-surface organisms.

Procedure Summary and Timeline

Prior to Week 1
- Identify four existing agroecosystems to serve as study sites; collect materials.

Week 1
- Set up pitfall traps (w/o soapy water).

48 h Prior to Day of Week 2
- Add soapy water to pitfall traps.

Week 2
- Identify and count captured organisms.

After Data Collection
- Analyze data and write up results.

Timing Factors

This investigation must be carried out when invertebrate fauna will be active in the agroecosystems selected for study.

Coordination with Other Investigations

The three different plots of Investigation 19 can be used as three study sites for this investigation.

Materials, Equipment, and Facilities

2 in. diameter soil corer

6+ m of 2 in. diameter PVC pipe for making pitfall traps

Forty plastic cups with lip diameter slightly larger than 2 in. (they should be able to fit inside the PVC with their lips resting on the end of the PVC)

Liquid detergent

Four plastic gallon jugs

Forty large zipper-top storage bags

Permanent marking pens

Plastic plates (if rainfall is likely during the study)

Screw-top vials (optional)

Tweezers

Dissecting needles

Dissecting microscopes or hand lenses

Large plastic petri dishes with lids (20 cm diameter)

Identification guides for terrestrial invertebrates likely to be found in local agroecosystems

Four different agroecosystems in which to place pitfall traps

Advance Preparation

- Locate four appropriate agroecosystems in which pitfall traps can be placed. The systems should all be in the same general vicinity and should vary as much as possible in cultural practices, crop type, and diversity. It would be ideal to include (1) a low-diversity, conventional monocrop system in which pesticides are used; (2) a high-diversity organic system incorporating some kind of mulch; (3) a cropping system directly adjacent to a natural or less disturbed ecosystem; and (4) a natural or less disturbed ecosystem, such as a meadow, woodland, grassland, or long-fallow field, near the other systems.
- Locate guides that can be used by students for the identification of organisms. If the instructor does not have a background in entomology, it will also be very useful to ask an entomologist to attend the lab meetings in which organisms will be identified.
- Collect the materials needed for construction of the pitfall traps.

Ongoing Maintenance

None required, but note that soapy water must be put into the plastic cups in the pitfall traps 48 h prior to the time students will collect the captured organisms. Students may be recruited to perform this task, but it may also fall upon the instructor or other staff.

Investigation Teams

Form four teams, each made up of three to five students. Each team will be responsible for censusing the soil-surface fauna of one agroecosystem.

Procedure

Setup

1. Meet as a group to decide where to place the 10 pitfall traps in your agroecosystem. You may place the traps randomly in the system (determined, e.g., by randomly generated x, y coordinates), along one or two transects within the system or according to some regular pattern.

2. Draw a map of your trap layout in your lab notebook.

3. One week before sampling is to begin, put the pitfall traps in place.

   a. Cut ten 15 cm long sections of 2 in. diameter PVC pipe. Make sure the cuts are square.

   b. Using a permanent marker, label each of the 10 plastic cups with an identifying number.

   c. At each trap location, use a 2 in. (5 cm) diameter soil corer to make a hole 15 cm deep.

   d. Push a PVC pipe section into each hole so that the end of the pipe is just below the surface of the soil.

   e. Place a plastic cup into each pipe so that its lip rests on the edge of the PVC pipe and is flushed with the surface of the soil. On the map of your trap layout, record the number of the cup at each trap location.

   f. If rainfall is possible during the next week, construct rain screens for the traps. These can be made of plastic plates propped up over each trap with a stick, piece of wood, or stiff wire.

4. If possible, check the pitfall traps every few days and release any organisms that have fallen into them (just remove the cup and dump out the organisms).

5. At some point before the traps are to be activated, make up a soap solution in a gallon jug by filling the jug and mixing in a few milliliters of liquid detergent.

   The following steps, in which traps are "activated," should be completed 24–48 h prior to the day captured organisms are to be removed from the traps for identification and counting.

6. Fill the bottom of each trap's cup with about a centimeter of the soap solution. The soap solution will prevent any arthropod that falls in the trap from crawling out (it may not prevent gastropods from crawling out).

7. Check the interface between the soil and each cup carefully, making sure there are no gaps.

Data Collection

1. After the traps have been active for 24–48 h, remove each cup and pour its contents into a zipper-top storage bag. Make sure that any very small organisms, which may resemble specks of dirt, make it into the bag. Mark each bag with the number of the cup whose contents were poured into the bag. Put each emptied cup back into its pipe and take the bags back to the lab.

2. In the lab, remove the trapped organisms from each bag of soapy water. Look carefully for very small organisms. Keep track of which bag (trap) each organism came from.

3. Separate the organisms from each bag into same-species piles. If desired, place each species into a separate vial.

4. With the aid of a dissecting scope (or hand lens) and available identification guides, identify the "morphospecies" of each organism—a taxonomic level that adequately distinguishes the species from others. (For many groups, identification beyond genus or even family is extremely difficult.) Ask the instructor, teaching assistant, or invited expert for assistance with identification and for verifications. As a general orientation to invertebrate taxonomy, Table 12.1 lists the groups of invertebrates that are commonly found in the soil-surface habitat—not all of which are easily captured in pitfall traps (or noticed if they are caught, because of their very small size).

5. In your lab notebook, record the morphospecies of each organism and its distinguishing characteristics. Make drawings if time allows. Give each morphospecies a unique name, even if it is as simple as "earthworm sp. #1."

6. For each species or morphospecies, count the number of individuals from each trap, and enter this figure in the Soil-Surface Fauna Census Datasheet (Figure 12.1).

7. Find the total number of individuals of each species or morphospecies captured in all the traps, and record this figure in the datasheet.

TABLE 12.1
Common Groups of Invertebrate Soil, Soil-Surface, and Litter Fauna

Phylum	Superclass or Class	Subclass or Order	Common Name
Platyhelminthes	Turbellaria		Flatworms
Aschelminthes	Nematoda		Roundworms
Rotifera			Rotifers
Mollusca	Gastropoda	Pulmonata	Snails and slugs
Annelida	Oligochaeta		Earthworms
Arthropoda	Crustacea	Isopoda	Woodlice and sow bugs
	Myriapoda	Diplopoda	Millipedes
		Chilopoda	Centipedes
	Insecta	Collembola	Springtails
		Isoptera	Termites
		Coleoptera	Beetles
		Diptera	Flies
		Lepidoptera	Moths, butterflies (as larvae)
		Hymenoptera	Ants, etc.
		Dermaptera	Earwigs
	Arachnida	Scorpionidae	Scorpions
		Araneida	Spiders
		Acarina	Mites

Soil-Surface Fauna Census Datasheet

Sampling Date: July 23, 2015

System: Organic Lettuce with Mulch

Organism	Number of Individuals in Each Trap										Total Individuals
	Trap 1	Trap 2	Trap 3	Trap 4	Trap 5	Trap 6	Trap 7	Trap 8	Trap 9	Trap 10	
Slug	2	3	1	0	0	2	4	1	0	3	16
Sow bug	8	3	4	1	3	6	4	0	2	1	32
Centipede	1	0	0	0	3	1	2	0	1	0	8
Earthworm	2	1	0	0	3	0	1	0	2	0	9
Ant sp. #1 (sm. black)	42	0	25	31	5	2	49	4	2	0	160
Ant sp. #2 (lrg. red)	2	0	0	0	4	0	0	22	0	0	28
Spider sp. #1	3	4	7	2	1	0	0	3	5	2	27
Spider sp. #2	1	0	0	0	0	1	0	0	0	0	2
Collembolid sp.	12	34	44	28	2	3	8	0	0	0	131

Figure 12.1
Examples of a Completed Soil-Surface Fauna Census Datasheet.

8. Using what you know about each organism's microhabitat, morphology, behavior, and coarse taxonomy, categorize each morphospecies into a "functional group." Functional groups include the following:

   a. Predators (spiders and some types of insects, such as carabid beetles)

   b. Parasites and parasitoids (e.g., small parasitic wasps)

   c. Detritus feeders (e.g., sow bugs)

    d. Scavengers (e.g., stink bugs)

    e. Omnivores (e.g., ants)

    f. Stem-boring herbivores

    g. Leaf-chewing herbivores (e.g., grasshoppers)

    h. Sucking herbivores (e.g., aphids)

    Herbivores are generally pests, and predators, parasites, and parasitoids are generally beneficials.

9. When you are certain that sampling is complete, remove the pitfall traps from the field and backfill the holes.

Data Analysis

1. Rank the various organisms by abundance. Calculate the relative abundance of each species/morphospecies (number of individuals of species/total number of individuals).

2. Determine if there was any relationship between trap location and type of organism captured.

3. Estimate the size of each population of organism encountered during your sampling.

4. Share your team's data with the other teams, and gather other teams' data.

5. For each species, compare the number of individuals captured in each agroecosystem.

6. Compare the number of species encountered in each agroecosystem.

7. *Optional*: Calculate a Shannon Diversity Index for the terrestrial invertebrates in each system. (See Investigation 7 or 10 for descriptions of how to calculate this index.)

Write-Up

The following are some suggestions for reporting on the results of the investigation:

- Describe and discuss any patterns of population distribution indicated by your sampling.

- Compare each agroecosystem as a habitat for soil-surface organisms. How does species diversity compare across systems? How do population sizes compare? The ratio of beneficials to pests? How do management practices appear to impact the populations?

- For each major species of terrestrial invertebrate encountered, discuss the factors that seem to determine the size of its population and its presence or absence in a cropping system.

- Discuss the probable bias associated with the pitfall trap method of sampling. What types of organisms are more likely and less likely to be detected by this method?

- Discuss the principal types of variation in terrestrial invertebrate populations (spatial, temporal) and ways your sampling did or did not account for them.

- Discuss the validity of making population estimates of each organism based on your sampling methods.

- Discuss how the time of year may have influenced the populations of soil-surface organisms in the systems studied.

- Based on your categorization of the organisms into functional groups, construct a hypothesized food web for each agroecosystem.

Variations and Further Study

1. Choose agroecosystems to study that border less disturbed or natural ecosystems, and set up the pitfall traps along transects at varying distances from the border. This setup will allow some inferences to be made about the effect of the neighboring system on the populations encountered in each agroecosystem.

2. Leave the traps in place over an entire cropping season, and "activate" them periodically to study changes in populations and diversity over time. When traps are not active, cover them with some kind of lid or cap to prevent organisms from being trapped.

3. Construct various types of pitfall traps to see how they vary in effectiveness with different types of organisms. Larger traps, traps containing different liquids, traps containing bait (e.g., tuna fish or sugar), and traps disguised as dark retreats are all possibilities.

Soil-Surface Fauna Census Datasheet											
Sampling Date:											
System:											
Organism	Number of Individuals in Each Trap										Total Individuals
	Trap 1	Trap 2	Trap 3	Trap 4	Trap 5	Trap 6	Trap 7	Trap 8	Trap 9	Trap 10	

Studies of Interspecific Interactions in Cropping Communities

Bioassay for Allelopathic Potential

Background

Allelopathy is an addition interference (Chapter 11 of *Agroecology*: *The Ecology of Sustainable Food Systems*), an interaction in which a plant adds some compound to its immediate environment, which has some effect on the growth, survival, or reproduction of other plants. Allelopathy is quite different from competition, which involves an organism removing a limited resource from the environment.

Allelopathic interactions have been shown to be at work in the inhibition of (1) crops by certain weeds, (2) weeds by certain cover crops, and even (3) weeds by certain crops (Chapter 11 of *Agroecology*: *The Ecology of Sustainable Food Systems*). Allelopathic stimulation of growth has also been noted. Cover crops and crops with allelopathic effects can help in the management of weeds in agroecosystems and thus play a role in reducing the use of externally derived synthetic inputs.

Determining if allelopathic inhibition (or stimulation) is actually occurring in a system is a multistep process (Chapter 11 of *Agroecology*: *The Ecology of Sustainable Food Systems*). In the first phase, the investigator determines if an extract of a plant has the potential to affect the growth of other plants. This work is performed with a technique known as a bioassay. In the several steps that make up the second phase, the investigator demonstrates allelopathic inhibition in the field and may try to isolate the suspected allelopathic compound.

Textbook Correlation

Investigation 11: Biotic Factors (Allelopathic Modification of the Environment)

Investigation 16: Species Interactions in Crop Communities (Interference at the Community Level)

Synopsis

Using a rapid bioassay technique, water extracts of a selected weed species and a selected crop species are tested in the lab for their potential to allelopathically inhibit the seed germination and seedling development of four different crop species, two of which are closely related to common weeds.

Objectives

- Investigate the phenomenon of allelopathy.
- Learn the rapid bioassay technique for determining allelopathic potential.

- Practice precision laboratory skills.
- Determine if two selected plant species commonly present in agroecosystems may possess allelopathic potential.
- Use a student's *t*-test and a chi-square test to determine the statistical significance of data.
- Discuss the implications of the results for the management of agroecosystems.

Procedure Summary and Timeline

Prior to class time

- Collect materials; prepare leachate solutions.

Week 1

- Set up the bioassay chambers.

Week 2

- Collect data.

Following week 2

- Analyze data and write up results.

Timing Factors

Even though this investigation is performed in the lab, it requires that healthy crop and weed material be available for making water extracts (leachates). The best results are obtained when this plant material has not recently been rained on or irrigated. Because it needs little advance preparation and only two lab section meetings, this investigation can be worked in between other investigations of greater length and logistical complexity.

Materials, Equipment, and Facilities

Fifteen petri dishes per team (= 60 total)

3–5 kg of clean, sterile, uniformly screened sand

Thirty #1 qualitative 9 cm diameter filter paper discs per team (=120 total)

Crop plant material (to be tested for allelopathic potential)

Weed plant material (to be tested for allelopathic potential)

Seeds of two crop species (e.g., corn, beans, squash, and broccoli)

Seeds of two agricultural species closely related to common weeds (e.g., radish, oats, and vetch; these are hereafter referred to as "weed-like" species)

Twenty 50 mL beakers

Twenty 10 mL volumetric pipettes with pipetters

Four quart–size jars (for soaking plant material to make leachate)

Vacuum filter

Four 250 mL beakers (for holding leachate)

Tape suitable for labeling

Label marker

Tweezers

Parafilm

Incubator for seed germination

Refrigerator

Metric rulers

Distilled water

Five copies of the Bioassay Datasheet, per student

Five copies of the Bioassay Sum of Squares (SS) Worksheet, per student

Two copies of the Bioassay *t*-Test Worksheet, per student

Two copies of the Bioassay Chi-Square Worksheet, per student

Advance Preparation

1. Select the crop and the weed to be tested for allelopathic potential. Good crop candidates include squash, beans, corn, and tomato. Good weed candidates include wild mustard (*Brassica campestris*), redroot pigweed (*Amaranthus retroflexus*), lambsquarters (*Chenopodium album*), and wild oats (*Avena fatua*).

2. Select two crop species and two weed-like species on which the testing will be conducted (these are referred to hereafter as "indicator species"). Secure supplies of seed of each species (about 250 seeds each).

3. Test the viability of the seeds of the indicator species by sowing about 10 of each in wet sand in flats and placing the flats in a warm area for a week. Do not use seeds with a germination rate below about 75%. (However, if using seed collected from wild or weed species, a germination rate as low as 25% is acceptable.)

4. Secure a supply of good-quality sand. It is very important that the sand be washed, free of organic matter, free of salt contamination, sterile, dry, and of uniform grain size. Washed, screened, and kiln-dried commercial sand is best.

5. Collect all other materials.

6. Prepare leachate solutions of the selected weed and the crop plant. The soaking process can be initiated just 1 h or so prior to the beginning of the lab section and completed while students set up their bioassay chambers. The following steps detail the procedure for creating a leachate solution for one species of plant:

    a. Collect plant material in the field, as long as possible after rain or irrigation.

    b. Dry the plant material in an area with good air circulation for approximately 96 h (longer if the relative humidity is high).

    c. Weigh out 2.5 g of dry plant material, and place it in a quart jar or 1 L beaker with 100 mL of distilled water. Soak the mixture for 2 h at approximately 25°C. Filter the extract under vacuum to obtain "2.5%" leachate.

    d. Place the filtered extract in a 250 mL beaker for common use by investigation teams.

    e. Repeat steps c and d with 5 g of dry plant material to obtain "5%" leachate.

7. Photocopy the appropriate number of datasheets and worksheets.

Ongoing Maintenance

After 72 h of incubation, the seeds have germinated, and their roots and shoots are ready to be measured. Since most lab sections meet on a weekly basis, it will generally be necessary to move the bioassay chambers to a refrigerator (to arrest growth) after 72 h.

Investigation Teams

Form four teams, three to six students per team. Each team will test the effects of two concentrations of the weed extract and two concentrations of the crop extract on one indicator species, either a crop or a weed-like crop. There will also be a distilled-water control treatment, for a total of five different treatments per indicator species. Each treatment will have three replicates. See the following logistics map (if there are enough students for a fifth team, that group can test the leachates on an additional crop seed or weed-like seed):

Team 1	Team 2	Team 3	Team 4
Weed-Like Seed A	Weed-Like Seed B	Crop Seed A	Crop Seed B
5 treatments × 3 replicates = 15 bioassay chambers	5 treatments × 3 replicates = 15 bioassay chambers	5 treatments × 3 replicates = 15 bioassay chambers	5 treatments × 3 replicates = 15 bioassay chambers

Procedure

Before beginning the setup, make sure that each team has selected or been assigned a seed type (as an indicator species).

Setup

The following steps detail the procedure for one set of bioassay chambers, in which one indicator species (in seed form) is used to test the different leachate solutions:

1. Prepare the seeds.
   a. Obtain 160–200 seeds of the indicator species assigned to the team.
   b. Distribute the seeds evenly among five 50 mL beakers. To each beaker, add enough of one of the following solutions to cover the seeds: distilled water, 2.5% weed leachate, 5% weed leachate, 2.5% crop leachate, and 5% crop leachate. Label the beakers appropriately.
   c. Allow the seeds to soak for 1 h. In the meantime, begin step 2.
2. Prepare the bioassay chambers.
   a. Obtain 15 petri dishes.
   b. Weigh out 45 g of sand into each petri dish. Level the sand.
   c. Place a filter paper disc on the sand surface in each dish.
   d. Label the lid of each dish with date, team, indicator species, treatment type, and a subscript to indicate replicate number (Figure 13.1). Lay out the petri dishes on the table surface as shown in Figure 13.1.
3. Irrigate the chambers.
   a. Using a volumetric pipette with "pipetter," add 10 mL of distilled water to each of the three control chambers.
   b. Using a clean pipette, add 10 mL of 2.5% weed leachate to each of the three appropriate chambers.
   c. Using a clean pipette, add 10 mL of 5% weed leachate to the appropriate chambers.
   d. Repeat steps 3b and c for the two concentrations of crop plant leachate.

   Note: If there is a shortage of pipettes, you can get by with two pipettes per team. Use one for the control solution, 2.5% weed leachate, and 5% weed leachate (in that order) and another for the 2.5% crop plant leachate and 5% crop plant leachate (in that order).

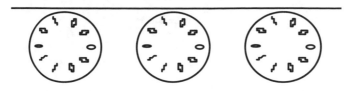

Figure 13.1
Setup of one team's bioassay chambers.

4. Add seeds to the chambers.

   a. Using tweezers, place 10 seeds onto the surface of each chamber.

   b. Arrange the seeds in a circle near the outside edge of the filter paper. If the hypocotyl (root) and epicotyl (shoot) ends of the seed can be distinguished, point the hypocotyl end of each seed toward the middle of the chamber.

5. Seal the chambers.

   a. Soak three filter paper discs in distilled water, and use them to cover the seeds in each of the three control chambers. Repeat this step for the other treatments, using the appropriate leachate solution for each set of chambers.

   b. Place a petri dish lid on each of the 15 chambers.

   c. Seal each chamber with parafilm.

   d. Make sure the chambers are labeled appropriately, and place them in a dark incubator at 25°C–27°C for 72 h. (Move to a refrigerator after 72 h to arrest growth until data can be taken.)

Data Collection

The following steps describe the data collection procedures for one set of bioassay chambers (those of one team, all with the same indicator species):

1. Open each chamber of the control treatment and measure the length of the hypocotyl (root) and epicotyl (shoot) of each germinated seed to the nearest millimeter. Record the data on a copy of the Bioassay Datasheet (Figure 13.2) (alternatively, or in addition, set up an electronic spreadsheet to enter the data; this spreadsheet can also be used for data analysis, as described later). If there is no measurable root or shoot, enter a dash instead of 0. When the data are analyzed for statistical significance, such observations are not included.

2. Find the total root length and the total shoot length for each chamber. Transfer the totals to the lower chart of the datasheet. Calculate a combined total for roots and a combined total for shoots and enter these totals in the lower chart.

3. Determine the *n* for the root data. This is the number of observations greater than 0 in all three chambers. Do the same for the shoot data. Note that the *n* for roots may be different from the *n* for shoots.

4. Divide the combined total for roots by the *n* for roots to calculate the mean root length. Enter this figure in the chart. Do the same for the shoot data.

5. Count the number of germinated seeds in each chamber. These are the seeds that produced a measurable root, a measurable shoot, or both. Record these figures on the datasheet and calculate a total for the treatment. *Note that the number of germinated seeds may differ from both the* n *used to calculate the root length mean and the* n *used to calculate the shoot length mean.*

6. Repeat steps 1–5 for each of the other four treatments, using a different copy of the datasheet for each treatment.

Bioassay Datasheet

Sampling Date: November 23, 2015

Treatment: 2.5% *Chenopodium* Extract

Indicator Species: Squash

Chamber A			Chamber B			Chamber C		
Seed	Root Length	Shoot Length	Seed	Root Length	Shoot Length	Seed	Root Length	Shoot Length
1	7	—	1	—	—	1	21	2
2	14	—	2	21	4	2	40	7
3	—	—	3	30	4	3	27	4
4	7	—	4	27	3	4	20	2
5	17	3	5	23	5	5	37	5
6	32	5	6	27	5	6	35	5
7	30	6	7	30	6	7	35	7
8	19	3	8	25	4	8	—	—
9	32	5	9	30	4	9	10	—
10	14	3	10	18	5	10	18	—
Total	172	25	Total	231	40	Total	243	32

	Root Length	Shoot Length	Number of Germinated Seeds
Total, chamber A	172 mm	25 mm	9
Total, chamber B	231 mm	40 mm	9
Total, chamber C	243 mm	32 mm	9
Combined total	646 mm	97 mm	27
Number of observations > 0 (=n)	27	22	
Mean (combined total/n)	23.93 mm	4.41 mm	

Figure 13.2
Example of a Completed Bioassay Datasheet.

Data Analysis

The data on root length and shoot length must now be analyzed to determine if the leachate treatments (1) reduced (or enhanced) root and shoot growth relative to the control and/or (2) reduced (or increased) the number of seeds that germinated. Comparison of experimental data with control data will show differences, but we need to know if these differences are due to random chance or allelopathic effects. A student's t-test will allow you to determine the likelihood of allelopathic inhibition (or stimulation) in the case of the root and shoot length, and a chi-square test will do the same for seed germination. The following steps listed describe the statistical analysis for one set of bioassay chambers (one team's data). Each team will share its analyzed data with the lab section as a whole.

1. Calculate the sum of squares (SS) for the root and shoot growth data of each treatment. The Bioassay SS Worksheet is provided for this purpose (you will need a copy for each treatment). A sample of a completed worksheet is shown in Figure 13.3. Calculating the SS is a necessary step in calculating the standard deviation for each set of data; the standard deviation is a useful statistic and is necessary for completing the t-test:

    a. Enter the figures for root and shoot length from the datasheet for the treatment being analyzed. Seeds 1–10 are those in chamber A, seeds 11–20 are those in chamber B, and seeds 21–30 are those in chamber C. For seeds without a measurable root and/or shoot, enter a dash.

						Bioassay Sum of Squares Worksheet			

Treatment: 2.5% *Chenopodium* Extract   Indicator Species: Squash

Seed	Root Length (x_i)	Mean Length ($\bar{x}$)	$x_i - \bar{x}$	$(x_i - \bar{x})^2$	Shoot Length (x_i)	Mean Length ($\bar{x}$)	$x_i - \bar{x}$	$(x_i - \bar{x})^2$
1	7	23.93	−16.93	286.6249				
2	14	23.93	−9.93	98.6049				
3	—	—	—	—				
4	7	23.93	−16.93	286.62				
5	17	23.93	−6.93	48.02				
6	32	23.93	8.07	65.12				
7	30	23.93	6.07	36.84				
8	19	23.93	−4.93	24.30				
9	32	23.93	8.07	65.12				
10	14	23.93	−9.93	98.60				
11	—	—	—	—				
12	21	23.93	−2.93	8.58				
13	30	23.93	6.07	36.84				
14	27	23.93	3.07	9.42				
15	23	23.93	−0.93	0.86				
16	27	23.93	3.07	9.42				
17	30	23.93	6.07	36.84				
18	25	23.93	1.07	1.14				
19	30	23.93	6.07	36.84				
20	18	23.93	−5.93	35.16				
21	21	23.93	−2.93	8.58				
22	40	23.93	16.07	258.24				
23	27	23.93	3.07	9.42				
24	20	23.93	−3.93	15.44				
25	37	23.93	13.07	170.82				
26	35	23.93	11.07	122.54				
27	35	23.93	11.07	122.54				
28	—	—	—	—				
29	10	23.93	−13.93	194.04				
30	18	23.93	−5.93	35.16				
			$SS = \sum(x_i - \bar{x})^2$	2121.85			$SS = \sum(x_i - \bar{x})^2$	
		Number of observations (n) =		27				

Figure 13.3
Example of a Partially Completed Bioassay SS Worksheet.

b. In each row of the mean ($\bar{x}$) column for the root data, enter the root length mean calculated on the datasheet. Do the same for the shoot data.

c. Find the difference between each observation and the mean, and square each result. Enter the figures in the appropriate columns. Do not calculate differences and squares for observations less than 1; leave those spaces blank.

d. Sum each of the squares columns to determine the SS for each set of data.

e. Repeat steps a–d for each of the remaining treatments, using a new copy of the worksheet for each.

Alternative method: If an electronic spreadsheet was used for recording the data and calculating means, it can also be used to calculate the sums of squares. In addition, if the spreadsheet software has a standard deviation function, the SS calculation can be skipped entirely, since its sole purpose is to calculate the standard deviation.

2. Calculate the standard deviation for each set of data. The standard deviation is a measure of how "spread out" the data points are in a set of data.

   a. Transfer the figures for *n*, the mean, and the SS for each set of data to copies of the Bioassay *t*-Test Worksheet (Figure 13.4). (You will need two copies of this worksheet, one for each leachate type.)

   b. Calculate the standard deviation for each set of data using the formula below:

$$S = \sqrt{\frac{\text{Sum of squares}}{n-1}}$$

   Enter the calculated *s* for each set of data on the *t*-test worksheet.

   Alternative method: If an electronic spreadsheet with a standard deviation function is being used, set up the spreadsheet to calculate the standard deviations.

3. Calculate a *t* value for each set of experimental data and enter it in the appropriate copy of worksheet. This value, called t_{calc}, will be compared with a standard value, t_{crit}, to determine if a set of experimental data is significantly different from the control data. The formula for t_{calc} is shown on the worksheet. The subscript 1 refers to data from the control treatment, and the subscript 2 refers to data from the experimental treatment.

Bioassay *t*-Test Worksheet

Leachate Species: *Chenopodium*

Indicator Species: Squash

	Treatment					
	Control		2.5% Leachate		5% Leachate	
	Root	Shoot	Root	Shoot	Root	Shoot
n	29	24	27	22	11	0
$\bar{x}$	30.83	3.88	23.93	4.41	2.18	0
SS	2548.14	44.63	2121.85	41.32	23.61	0
s	9.54	1.39	9.03	1.40	1.54	0
t_{calc}			3.89	−1.81	17.41	0
t_{crit}			1.68	1.68	1.69	1.70
Significant?			Yes	Yes	Yes	No

n, number of observations > 0

$(\bar{x})$, mean = $\Sigma x_i / n$

SS, sum of squares = $\sum (x_i - \bar{x})^2$

s, standard deviation = $\sqrt{\dfrac{\text{Sum of squares}}{n-1}}$

t_{calc}, calculated *t* value = $\dfrac{\bar{x}_1 - \bar{x}_2}{\sqrt{(n_1 - 1)s_1^2 + (n_2 - 1)s_2^2}} \left(\sqrt{\dfrac{n_1 n_2 (n_1 + n_2 - 2)}{n_1 + n_2}} \right)$

t_{crit}, critical *t* value = value determined in Data Analysis step 4

Figure 13.4
Example of a Completed Bioassay *t*-Test Worksheet.

If an electronic spreadsheet is being used, the formula for t_{calc} can be entered and the value calculated automatically. The data for the control treatment have no t-values because they are what the experimental data are being compared against.

4. Determine the appropriate value of t_{crit} for each set of experimental data. This value depends on two factors: degrees of freedom and significance (or confidence) level. The degrees of freedom are equal to $n_1 + n_2 - 2$ (n_1 is the n for the control data, and n_2 is the n for the experimental data). The significance level is the degree of confidence we have that the two sets of data are significantly different. We will use a significance level of 0.05, which sets the standard confidence interval for the comparison of means that we are making. For the purposes of this investigation, there are only a few likely values at a significance level of 0.05, as shown in the following table:

Degrees of Freedom ($n_1 + n_2 - 2$)	t_{crit} at 0.05 Significance Level
18–19	1.73
20–22	1.72
23–26	1.71
27–34	1.70
35–39	1.69
40–49	1.68
50–69	1.68

5. Compare the values of t_{crit} and t_{calc} for each set of experimental data to determine significance. If the absolute value of t_{calc} is greater than or equal to the value of t_{crit}, we can be highly confident that the two sets of data represent significant differences or, in other words, that the leachate solution inhibited or stimulated the growth of the indicator seedlings.

6. Determine if the germination rate for each experimental treatment is significantly different from the germination rate for the control treatment. We used a t-test to test for significant differences in root and shoot growth because these data are *continuous*; in contrast, the germination data are "discrete," which means we must use a chi-square (χ^2) test on them. The chi-square test produces a calculated chi-square value that is compared against a critical chi-square value to determine significance. Use copies of the Bioassay Chi-Square Test Worksheet to perform this test on the data (one copy for each leachate type (Figure 13.5)). Alternatively, set up an electronic spreadsheet to record the data and perform the calculations. The following steps describe the procedure for analyzing the germination rate of the seeds exposed to one of the leachate types:

a. Enter the germination data for the 2.5% treatment and the control treatment in the upper table of the worksheet (these data were recorded on datasheets during the data collection phase of this investigation). The subscripts of the variable names in each box refer to the row number and column number of the table (e.g., f_{12} is the value for row 1, column 2).

b. Sum the columns and the rows of the table to derive values for C_1, C_2, R_1, and R_2.

c. Calculate $|f_{11}f_{22} - f_{12}f_{21}|$ and $n/2$ and enter these figures in the table at the bottom of the worksheet.

d. Choose the appropriate chi-square formula to use, based on the results of step c.

e. Calculate the chi-square value using the appropriate formula, and enter the result in the bottom table.

f. Enter the critical chi-square value in the bottom table. In this two-row by two-column comparison, there is one degree of freedom. At a significance level of 0.05, one degree of freedom indicates a critical value of 3.841.

g. Compare the calculated chi-square value with the critical chi-square value. There is a significant difference in the germination rate of the two treatments if the calculated chi-square value is greater than the critical value.

h. Repeat steps a–g for the 5% treatment.

Bioassay Chi-Square Worksheet			
Leachate Species: *Chenopodium*		Indicator Species: Squash	
	Number of Germinated Seeds	**Number of Ungerminated Seeds**	**Total**
2.5% leachate treatment			
Leachate treatment	27 $(=f_{11})$	3 $(=f_{12})$	30 $(=R_1)$
Control treatment	29 $(=f_{21})$	1 $(=f_{22})$	30 $(=R_2)$
Total	56 $(=C_1)$	4 $(=C_2)$	60 $(=n)$
5% leachate treatment			
Leachate treatment	11 $(=f_{11})$	19 $(=f_{12})$	30 $(=R_1)$
Control treatment	29 $(=f_{21})$	1 $(=f_{22})$	30 $(=R_2)$
Total	40 $(=C_1)$	20 $(=C_2)$	60 $(=n)$
	2.5% Treatment	5% Treatment	
$\lvert f_{11}f_{22}-f_{12}f_{21}\rvert$	60	540	
$n/2$	30	30	
χ^2_{calc}	0.27	21.68	
χ^2_{crit}	3.841	3.841	
Significant?	No	Yes	

Formulas:

If $\lvert f_{11}f_{22}-f_{12}f_{21}\rvert \le n/2$, use this formula for χ^2_{calc}: $\chi^2 = \dfrac{n\left(f_{11}f_{22}-f_{12}f_{21}\right)^2}{C_1 C_2 R_1 R_2}$

If $\lvert f_{11}f_{22}-f_{12}f_{21}\rvert > n/2$, use this formula for χ^2_{calc}: $\chi^2 = \dfrac{n\left(\lvert f_{11}f_{22}-f_{12}f_{21}\rvert - n/2\right)^2}{C_1 C_2 R_1 R_2}$

Figure 13.5
Example of a Completed Bioassay Chi-Square Test Worksheet.

Write-Up

The data collected in this investigation are very rich, allowing many kinds of comparisons and analyses. When teams share their data, the class as a whole has available data showing the effects of the leachate of a crop species and a weed species (at two concentrations each) on the root growth, shoot growth, and seed germination of four species of plants. The following questions suggest some of the possible avenues of analysis:

- Which had the most significant effects, the crop-plant leachate or the weed leachate?
- Which seeds and seedlings were inhibited the most, those of the crop plants or those of the weed-like crops?
- What was the relative vulnerability to allelopathic inhibition among the four indicator species?
- Did the 5% leachate solutions differ in their effects in any systematic way from the 2.5% leachate solutions?
- What is the relationship between root vs. shoot inhibition and inhibition of germination?
- Were there any patterns in the way shoots responded to the treatments as compared to roots?
- Did inhibition occur more frequently than stimulation? If stimulation occurred, what do the cases of stimulation have in common?

Because of the complexity of this investigation's data, lab reports can be team written, with each member taking responsibility for a certain part of the report. All members, however, should contribute equally to discussion of the results and conclusions.

The following are some additional suggestions for writing up the results of the investigation:

- Present graphs or tables of data summarizing overall class results, and discuss the patterns that emerge from the results. Develop a method of indicating which experimental treatment means are significantly different from the control mean.
- Present graphs or tables showing selected patterns, comparisons, or relationships. In some cases, it will be useful to transform the data into a form that shows the desired patterns most clearly. For example, it might be useful to present a bar graph showing how the two leachate concentrations of one of the leachate species affected the root and/or shoot growth of each of the indicator species. This graph could be based on a measure of "percent inhibition," calculated for each set of data by expressing the following measure of inhibition as a percentage:

$$1 - \frac{\text{Mean root or shoot length of leachate treatment}}{\text{Mean root or shoot length of control treatment}}$$

This type of presentation of data can be used as a way of marshaling evidence to support particular conclusions or hypotheses.

- Propose explanations of the significant findings, in ecological or evolutionary terms.
- Discuss the implications of the findings for the management of agroecosystems.
- Discuss the limits of the investigation.
- Propose further study for testing of new hypotheses or exploration of the questions raised by the investigation.

Variations and Further Study

1. Test additional crop and weed species for allelopathic potential (e.g., if there is more than one lab section, each section can test a different set of species).
2. Conduct the bioassay tests on additional indicator species (e.g., if there are enough students in the lab section to form more than four groups, the additional groups can perform their bioassays on different indicator species).
3. Collect the seeds of actual weeds and use them in the investigation in place of the weed-like seeds.
4. Test the allelopathic potential of two-species mixtures of leachates (two weeds, two crops, or a crop and a weed) to determine if a mixture has a synergistic or complementary effect.
5. Test to see if the osmotic concentration of the leachate extracts may be a factor inhibiting the germination of the seeds or the growth of the seedlings. If a leachate extract is relatively concentrated, it can destabilize the osmotic balance of the seed, making it difficult to separate this effect from that of the chemical toxicity of the extract. The leachate extracts used in this investigation are not very likely to be concentrated enough to affect the seeds' osmotic balance, but it is a good idea to determine if there is an effect. This is done by using an osmometer to determine the solution strength of the two leachate solutions and then making up osmotically identical controls for each using the nontoxic sugar mannitol.

Bioassay Datasheet

Sampling Date:

Treatment:

Indicator Species:

	Chamber A			Chamber B			Chamber C	
Seed	Root Length	Shoot Length	Seed	Root Length	Shoot Length	Seed	Root Length	Shoot Length
1			1					
2			2					
3			3					
4			4					
5			5					
6			6					
7			7					
8			8					
9			9					
10			10					
Total			Total					

	Root Length	Shoot Length	Number of Germinated Seeds
Total, chamber A			
Total, chamber B			
Total, chamber C			
Combined total			
Number of observations > 0 ($= n$)			
Mean (combined total/n)			

Bioassay Sum of Square Worksheet

Treatment: _____　　　　　　　　　　　　　　Indicator Species: _____

Seed	Root Length (x_i)	Mean Length ($\bar{x}$)	$x_i - \bar{x}$	$(x_i - \bar{x})^2$	Shoot Length (x_i)	Mean Length ($\bar{x}$)	$x_i - \bar{x}$	$(x_i - \bar{x})^2$
1								
2								
3								
4								
5								
6								
7								
8								
9								
10								
11								
12								
13								
14								
15								
16								
17								
18								
19								
20								
21								
22								
23								
24								
25								
26								
27								
28								
29								
30								
			$SS = \sum \left(x_i - \bar{x} \right)^2$				$SS = \sum \left(x_i - \bar{x} \right)^2$	
		Number of observations (n) =						

Bioassay *t*-Test Worksheet						
Leachate Species:						
Indicator Species:						
	Treatment					
	Control		**2.5% Leachate**		**5% Leachate**	
	Root	**Shoot**	**Root**	**Shoot**	**Root**	**Shoot**
n						
$\bar{x}$						
SS						
s						
t_{calc}						
t_{crit}						
Significant?						

n, number of observations > 0

$(\bar{x})$, mean $= \Sigma\, x_i/n$

SS, sum of squares $= \sum \left(x_i - \bar{x} \right)^2$

s, standard deviation $= \sqrt{\dfrac{\text{Sum of squares}}{n-1}}$

t_{calc}, calculated t value $= \dfrac{\bar{x}_1 - \bar{x}_2}{\sqrt{\left(n_1 - 1\right)s_1^2 + \left(n_2 - 1\right)s_2^2}} \left(\sqrt{\dfrac{n_1 n_2 \left(n_1 + n_2 - 2\right)}{n_1 + n_2}} \right)$

t_{crit}, critical t value = value determined in Data Analysis step 4

	Bioassay Chi-Square Worksheet		
Leachate Species:		Indicator Species:	
	Number of Germinated Seeds	**Number of Ungerminated Seeds**	**Total**
2.5% leachate treatment			
Leachate treatment	$(=f_{11})$	$(=f_{12})$	$(=R_1)$
Control treatment	$(=f_{21})$	$(=f_{22})$	$(=R_2)$
Total	$(=C_1)$	$(=C_2)$	$(=n)$
5% leachate treatment			
Leachate treatment	$(=f_{11})$	$(=f_{12})$	$(=R_1)$
Control treatment	$(=f_{21})$	$(=f_{22})$	$(=R_2)$
Total	$(=C_1)$	$(=C_2)$	$(=n)$
	2.5% Treatment	5% Treatment	
$|f_{11}f_{22} - f_{12}f_{21}|$			
$n/2$			
χ^2_{calc}			
χ^2_{crit}			
Significant?			

Formulas:

If $|f_{11}f_{22} - f_{12}f_{21}| \le n/2$, use this formula for χ^2_{calc}: $\chi^2 = \dfrac{n\left(f_{11}f_{22} - f_{12}f_{21}\right)^2}{C_1 C_2 R_1 R_2}$

If $|f_{11}f_{22} - f_{12}f_{21}| > n/2$, use this formula for χ^2_{calc}: $\chi^2 = \dfrac{n\left(|f_{11}f_{22} - f_{12}f_{21}| - n/2\right)^2}{C_1 C_2 R_1 R_2}$

14

Rhizobium Nodulation in Legumes

Background

The mutualistic relationship between bacteria in the genus *Rhizobium* and plants in the Fabaceae family is one of the most important interspecific interactions in agroecosystems. As a result of this mutualism, nitrogen derived from the atmosphere is made available to all the biotic members of an agroecosystem and ultimately to the human managers of the system. This interaction demonstrates the value of relationships in which both members benefit (Chapter 16 of *Agroecology: The Ecology of Sustainable Food Systems*) and shows how such relationships contribute to the overall diversity and stability of an agroecosystem (Chapter 17 of *Agroecology: The Ecology of Sustainable Food Systems*).

Although the *Rhizobium* bacteria can survive as free-living organisms in the soil, they are much more successful when they infect the root of a legume plant and form the characteristic root nodule in which they live. Leguminous plants can likewise survive without nodules in their roots, but they develop much more strongly when the bacteria are resident. Thus, both members of the mutualism do much better when in association. In most natural ecosystems, naturally occurring *Rhizobium*–legume mutualisms are a principal mechanism for moving nitrogen from the air into ecosystem biomass, after which this essential element can be recycled within the system.

In early agricultural systems, and existing traditional agroecosystems, biologically fixed nitrogen from the mutualism is still the primary source of nitrogen input. Planting legumes in rotations, as cover crops and as intercrops, are the methods used to enhance the entrance of natural sources of nitrogen. Organic farmers, although heavily dependent on composted crop residues and manures for much of their nitrogen needs, have also become increasingly aware of the importance of this mutualism. To decrease the growing dependence of conventional agriculture on synthetic, fossil fuel–based sources of nitrogen, and the frequent environmental contamination that can accompany their use, we need more thorough knowledge of how this mutualism functions and how it can be used to improve the sustainability of agriculture.

Textbook Correlation

Investigation 11: Biotic Factors (Addition Interferences: Symbioses)
Investigation 16: Species Interactions in Crop Communities
Investigation 17: Agroecosystem Diversity

Synopsis

Seeds of a legume are planted in pots in three types of soil: (1) heat sterilized to kill all possible nitrogen-fixing bacteria, (2) inoculated with *Rhizobium* bacteria, and (3) untreated. Observations of nodulation and plant development are made and correlated with the presence or absence of the bacteria in the soil.

Objectives

- Learn a methodology for extracting nodules from the soil–root system.
- Correlate nodulation with nitrogen fixation and plant growth.
- Investigate some of the dynamics of the *Rhizobium*–legume mutualism.

Procedure Summary and Timeline

Prior to week 1
- Select test legume; obtain seeds; collect, prepare, and heat-sterilize soil; obtain *Rhizobium* inoculant.

Week 1
- Set up pots; inoculate seed; plant seed.

Week 2
- Thin seedlings.
- Two or three times between week 2 and the final week, make measurements of plant development.

Week 7, 8, or 9
- Remove plants from soil; measure root and shoot development; quantify nodule presence and development.

Timing Factors

This experiment can be performed at any time of the year if a controlled-climate greenhouse is available. Otherwise, it should be done during the normal growing season.

Coordination with Other Investigations

See "Variations and Further Study" section for a description of how a version of this investigation can be integrated into Investigation 19.

Materials, Equipment, and Facilities

Sixty 1 gal pots (large enough to hold about 3 kg of dry soil)

150 seeds of a legume appropriate to the region

Rhizobium inoculant appropriate to the selected legume

Approximately 300 kg (moist weight) of local soil

Meter sticks

Greenhouse or lathhouse

1 mm mesh soil screen

Garden hose with high-pressure nozzle

Paper towels

Electric, box-type soil sterilizer

Tweezers

Plant clippers or pruning shears

Paper drying bags for shoots and roots

Small envelopes for drying nodules

Marking pens

Sixty pot tags

Drying oven

Airtight containers for storing sterilized soil

Three copies of the Nodulation Datasheet, per student

Advance Preparation

- Select a bush bean variety that grows well in the local region and is known to have a dependence on *Rhizobium* for best growth and yield. A short cycle bean is best for the purposes of this investigation. Obtain approximately 150 seeds of the bean.
- Obtain the recommended bacterial inoculant for the chosen legume. Most seed suppliers also provide the proper inoculant.
- Collect approximately 300 kg of soil (a little more if the soil is very wet) from an area that is known to be a good producer of beans; if possible, use soil from a specific location with a history of bean planting so as to ensure the presence of a resident inoculum. Air-dry the soil, screen it through a 2 mm screen, and store it in a sealed container.
- Sterilize approximately 150 kg of the dried soil in the soil sterilizer by heating the soil at 100°C for 24–48 h. Once sterilized, the soil should be stored in a clear, airtight container until used. (If a soil sterilizer is not available, a drying oven or autoclave can be used to sterilize the soil in smaller batches.)
- Obtain bench space in a controlled-climate greenhouse. If a greenhouse is not available, locate an outdoor area, either a bench in a lathhouse or a clean, clear area in a garden close to a source of water.
- Photocopy the appropriate number of datasheets.

Ongoing Maintenance

Pots will need to be checked regularly and watered. It is important to maintain constant moisture in the pots so as to not stress either the plants or the bacteria. When watering, add enough water to fully moisten the soil, but not so much that excessive water leaches out from the bottom of the pots.

Investigation Teams

Form four teams, each with three to five students. Each team will be responsible for monitoring and collecting data for five pots in each of the three treatments of the investigation (a total of 15 pots). The number of teams and the number of pots per team can be altered for very small or very large classes, but keep the total number of pots per treatment at about 20.

Procedure

The following steps describe the setup, data collection, and data analysis procedures for one team's set of 15 pots.

Setup

1. Obtain fifteen 1 gal pots and 30 seeds of the legume.

2. Make tags for the five pots of each treatment. Each tag should list treatment type (sterilized, untreated, sterilized/inoculated), pot number, team, and date.

3. Prepare the pots.

    a. Fill five pots with 3 kg of unsterilized soil each.

    b. Fill 10 pots with 3 kg of sterilized soil each. Be careful not to contaminate the sterilized soil with unsterilized soil. Clean tools and hands if working back and forth between treatments.

    c. Insert the appropriate tags into the soil of each pot. The five tags for the sterilized/inoculated treatment go in pots with sterilized soil.

4. Sow the seeds in the pots of the sterilized and untreated treatments. In each pot, plant two seeds approximately 1 cm deep and 2–3 cm apart near the center of the pot.

5. Inoculate 10 seeds with the inoculant, following the instructions that come with the inoculant. This usually means that the seed is dipped in some kind of a sticking agent, after which the seed is placed in the inoculant powder.

6. Sow the inoculated seeds in the pots of the sterilized/inoculated treatment, using the same technique as for the other treatments.

7. Place all 15 pots together on the bench of the greenhouse or lathhouse (or in the preselected garden area).

8. Water the pots uniformly, taking care to moisten the soil thoroughly without disturbing the seeds.

    The next step is performed a week after the seeds are sown.

9. Selectively remove one of the two bean seedlings from each pot, keeping the most vigorously growing one of the pair. If only one seedling emerged, then leave that one. Remove the selected seedling carefully by pulling it from the soil in a way that does not disturb the soil of the remaining seedling.

Collection of Plant Development Data

The following steps should be performed at least two times while the bean plants are growing and then again on the day the plants are dug up and final data collected. There is space on the datasheet for taking these data four total times for each treatment. The data on plant height and leaf number will be graphed to show development over time.

1. Measure the height of each plant (soil level to the tallest point of the plant) to the nearest 0.1 cm. Record these data on copies of the Legume Nodulation Datasheet (Figure 14.1) (use a separate datasheet for each treatment).

2. Count the number of leaves on each plant. A leaf counts as a leaf when it has fully opened and shows all three leaflets (it need not be full size). Do not count cotyledons. Record each plant's leaf number in the appropriate row of each datasheet.

Final Data Collection

The ideal time to collect data on *Rhizobium* nodulation (and final data on bean plant development) is at the time of maximum nodulation. Several studies have shown that maximum nodulation occurs when a bean plant's first pod is about 2.5 cm in length. So, it is advisable to monitor pod length and collect final data in the week that one or more pods reach 2.5 cm. If the time available in the course limits the length of the investigation, the final data can be collected sooner. Beans will reach the 2.5 cm pod stage of development

Legume Nodulation Datasheet								
Treatment: Untreated Soil				Indicator Species: Bush Bean				
	Pot/Plant Number							**Standard Deviation**
	1	**2**	**3**	**4**	**5**	**Total**	**Mean**	
Date: April 13 — Height (cm)	15	16	14	15	15	75	15	0.70
Number of leaves	2	3	2	3	2	12	2.4	0.55
Date: April 27 — Height (cm)	23	24	22	23	24	116	23.2	0.84
Number of leaves	5	6	4	5	5	25	5	0.71
Date: May 11 — Height (cm)	30	32	29	30	31	152	30.4	1.14
Number of leaves	8	9	7	8	8	40	8	0.71
Date: May 25 — Height (cm)	40	43	38	41	40	202	40.4	1.82
Number of leaves	15	16	14	15	15	75	15	0.71
Shoot mass (g)	15.1	18.4	14.8	15	14.9	78.2	15.6	1.55
Root mass (g)	3.3	5.1	3.2	3.4	3.2	18.2	3.6	0.82
Shoot/root ratio	4.6	3.6	4.6	4.4	4.7	21.9	4.4	0.44
Number of nodules	25	27	20	23	25	120	24	2.65
Mass of all nodules (mg)	8.75	9.34	8.05	8.53	8.80	43.47	8.69	0.47
Mean mass per nodule (mg)	0.35	0.34	0.40	0.37	0.35	1.82	0.36	0.02
Color of nodule interiors	Pink	Pink	Pink	Pink	Pink			

Figure 14.1
Example of a Completed Legume Nodulation Datasheet.

anywhere from 45 to 60 days after sowing, depending on the variety and the environmental conditions of the location:

1. Take the final measurements of plant development as described previously.
2. Harvest the aboveground plant material by cutting the stem at soil level. Put each plant in a marked paper bag and place the bag and plant in a drying oven to dry for 48 h at 70°C.
3. Separate each plant's root system from the soil in its pot, being careful not to dislodge any nodules from the roots.
   a. Carefully remove the soil–root mass from the pot. If it is hard and cemented (as might occur with heavy clay soils), place the soil–root mass into a bucket and soak for at least 15 min before moving on to the next step.
   b. Place the root–soil mass on 1 mm mesh soil screen (1 mm hardware cloth mounted on a wooden frame will do).
   c. Using a high-pressure nozzle and hose, carefully wash the soil from the root system, retrieving any roots or nodules that break off.
   d. When the root system is free of soil particles, pat it dry between paper towels.
4. Collect the nodules from each root system.
   a. Use tweezers to pick off each nodule.
   b. Count the number of nodules and record the number on the appropriate datasheet.
   c. Break open a few nodules and note the color inside. Red or pink implies active nitrogen fixation by the bacteria; white implies less activity or none at all.
   d. Place the nodules in a small labeled envelope, and place the envelope in a drying oven to dry for 48 h at 70°C.

5. Put the denoduled roots in a marked paper bag and place the bag in a drying oven to dry for 48 h at 70°C. Perform the following steps after the shoots, roots, and nodules have all been dried.

6. Find the mass of each dried shoot to the nearest 0.1 g and record the mass on the appropriate datasheet.

7. Find the mass of each dried root system to the nearest 0.1 g and record the mass on the appropriate datasheet.

8. Find the mass of each set of dried nodules to the nearest 0.01 mg and record the mass on the appropriate datasheet.

Data Analysis

1. Calculate the shoot/root ratio for each plant by dividing the dry mass of the shoot by the dry mass of the root system. Record the ratio on the datasheet.

2. Calculate the average mass per nodule for each plant by dividing the combined mass of the plant's nodules by the number of nodules.

3. Calculate the mean for each set (i.e., row) of data.

4. Calculate the standard deviation for each set of data (see Investigations 1, 3, 8, or 13 for a description of how to calculate this statistic).

5. Share your team's data with the other teams.

6. Gather the data from the other teams and find the overall means for the 20 plants in each treatment.

7. *Optional*: Perform appropriate statistical tests to determine if there are significant differences in plant growth and nodulation among the three treatments.

Write-Up

The following are some suggestions for reporting on the results of the investigation:

- Graph the growth and development data (number of leaves and height of plants) over time.
- Develop bar graphs to allow comparison of the measurements of plant development and nodulation among the three treatments.
- Discuss the differences that were observed, especially between the sterilized treatment and the other two. Did the presence of the bacteria make a difference in the performance of the bean plants? How did the sterilized/inoculated treatment compare to the unsterilized soil treatment?
- Discuss the implications of the results.

Note: In making inferences based on the nodulation data, take into account that the presence of nodules does not necessarily mean nitrogen fixation is actively taking place. Consider the results of the color test performed earlier when making inferences about nitrogen fixation.

Variations and Further Study

1. If the equipment is available, measure the nitrogen content of the plants and soil in the sterilized/inoculated and sterilized treatments. These data will allow a determination of how much nitrogen was added to each soil–plant "system" in the sterilized/inoculated treatment.

2. A field version of this investigation can be carried out by planting uninoculated bean seeds in two different soils that vary in their legume-cropping history. One location could be where beans or some other legumes were recently planted, ensuring the chance of good resident inocula, and the other location could be where legumes (even weedy legumes) have not been present for many years.

3. Test the hypothesis that when bean plants are in the vicinity of non-nitrogen-fixing but heavy-nitrogen-using plants (such as sweet corn), nodulation and nitrogen-fixing activity are stimulated compared to when the same beans are monocropped. An easy way to perform this test is to use a legume as one of the crops in Investigation 19 and to compare legume growth and nodulation in the monocrop legume plots and the legume–second crop inter-crop plots.

4. Certain crops, such as members of the Brassicaceae family, are known to inhibit the nodulation and activity of *Rhizobium*. Test the effect of intercropping beans with a brassica, using a monocrop of beans as the control.

5. In pots or in the field, sow inoculated bean seeds in soils with a range of levels of available nitrogen from soil amendments or fertilizer, and measure the effects of the nitrogen levels on nodulation. Excess nitrogen in the soil (especially highly water-soluble nitrogen from synthetic fertilizer) is known to inhibit nitrogen fixing by *Rhizobium*, either by preventing the bacteria from infecting legume roots or by inhibiting the bacteria's activity if it does infect the roots.

Legume Nodulation Datasheet									
Treatment:				Indicator Species:					
		Pot/Plant Number					Total	Mean	Standard Deviation
		1	2	3	4	5			
Date:	Height (cm)								
	Number of leaves								
Date:	Height (cm)								
	Number of leaves								
Date:	Height (cm)								
	Number of leaves								
Date:	Height (cm)								
	Number of leaves								
	Shoot mass (g)								
	Root mass (g)								
	Shoot/root ratio								
	Number of nodules								
	Mass of all nodules (mg)								
	Mean mass per nodule (mg)								
	Color of nodule interiors								

Investigation

Effects of Agroecosystem Diversity on Herbivore Activity*

Background

Herbivorous insects are one of a farmer's most serious antagonists. Most crop plants can withstand minor herbivory without ill effect, but more extensive herbivory can greatly reduce plant vitality, limit overall production, and lower the quality of the marketed crop.

Conventional monocultures have proven to be extremely susceptible to attack by herbivores. In the uniform conditions of a vast monoculture, with few organisms present besides an attractive crop plant, herbivore populations can increase exponentially and decimate the crop. Farmers often respond by applying synthetic chemical insecticides to control the herbivore populations, leading to a variety of other serious problems.

Alternative strategies for managing herbivore populations without using insecticides have been explored. Many alternative methods are based on the idea that an agroecosystem can be designed so that it manages herbivore populations by itself, with minimal external inputs being needed. The key characteristic of such an agroecosystem is higher diversity (Chapter 17 of *Agroecology: The Ecology of Sustainable Food Systems*).

The greater the diversity of an agroecosystem, the more likely it is to harbor beneficials that prey upon herbivores, the more likely it is to contain plants that chemically repel herbivores, and the more likely its more heterogeneous structure will disrupt the movement of herbivores or their search for target plants (Chapter 17 of *Agroecology: The Ecology of Sustainable Food Systems*). While the connection between higher diversity and less herbivory is well established as a general principle, its application in specific agroecosystems needs much more extensive testing. Diversity can be increased in a variety of ways (Chapter 17 of *Agroecology: The Ecology of Sustainable Food Systems*), and each will have different impacts depending on the crops involved, the unpredictable emergent qualities of the systems in which they are employed, and the presence of pests locally.

Textbook Correlation

Investigation 14: The Population Ecology of Agroecosystems (Applications of Niche Theory to Agriculture: Biological Control of Insect Pests)
Investigation 16: Species Interactions in Crop Communities
Investigation 17: Agroecosystem Diversity

---

* This investigation is designed to run in conjunction with Investigation 19. However, it is possible to adapt it to a different set of existing agroecosystems or to set up plantings on which to perform the comparisons described here (see "Variations and Further Study" section).

Synopsis

One of the crops grown in Investigation 19 is selected as the study crop. Herbivore consumption of study crop plants in the monoculture and polyculture treatments is measured, and the level of consumption found in the two systems is compared.

Objectives

- Learn techniques for measuring herbivore consumption of leaves.
- Relate agroecosystem diversity and structure to the incidence of herbivory.
- Test the hypothesis that a more diverse system will experience less herbivory than a less diverse system.

Procedure Summary and Timeline

Week 1
- Select leaves and measure consumption.

After week 1
- Analyze data and write up results.

Timing Factors

This investigation can be completed in just 1 or 2 weeks, at any point in the latter developmental stages of the crop plants in Investigation 19.

Materials, Equipment, and Facilities

Leaf area meter (recommended but not necessary; see Data Collection step 5 for substitute equipment)

Clear acetate plastic, in sheets or a roll (about 25–100 sheets, or the equivalent, will be needed, depending on the size of the leaves)

Permanent black markers, with fine tips

Random number generator or table of random numbers

Two copies of the Herbivory Comparison Datasheet, per student

Advance Preparation

- Collect necessary materials.
- Select a study crop from among the two crops growing in the plots used for Investigation 19. An ideal study crop is attacked readily by herbivorous insects and has leaves that are relatively flat and firm and of average size.
- Make the appropriate number of photocopies of the datasheet.

Ongoing Maintenance

No additional ongoing maintenance is required beyond that described for Investigation 19.

Investigation Teams

This investigation can be carried out by the same four teams formed for Investigation 19. Each team will be responsible for measuring herbivore consumption in its assigned block. A team will therefore measure consumption in one monoculture plot and one intercrop plot. See the following logistics map.

Team 1	Team 2	Team 3	Team 4
Block I	Block II	Block III	Block IV
24 leaves from monoculture plot; 24 leaves from intercrop plot	24 leaves from monoculture plot; 24 leaves from intercrop plot	24 leaves from monoculture plot; 24 leaves from intercrop plot	24 leaves from monoculture plot; 24 leaves from intercrop plot

Procedure

Data Collection

Each team will measure herbivore consumption in two plots in its assigned block: the monoculture plot of the study crop and the study-crop component of the intercrop plot. The following steps describe the procedure for measuring the herbivore consumption in one plot:

1. Generate a list of four random numbers, none of which is greater than the number of rows in the plot. These will be taken to the field and used to randomly select four rows in the plot. (Refer to the Introduction for information on generating random numbers.)

2. Generate four lists of random numbers, each list containing six numbers between 1 and 100 ($Q = 6$ and $M = 100$). These lists will be taken to the field and used to randomly select leaves from each selected row.

3. Sequentially number each row in the plot. Consult the list of four random numbers and mark the four rows whose numbers match those of the four random numbers.

4. Randomly select six leaves from each of the four marked rows.

   a. Identify one investigator as the "leaf counter" and one as the "observer".

   b. Choose one of the marked rows.

   c. Choose one of the four lists of six random numbers each.

   d. Beginning at one end of the chosen row, the leaf counter (1) arbitrarily selects a mature, fully developed leaf and calls out "one"; (2) selects another leaf from the same plant and calls out "two"; (3) moves on to the next plant, arbitrarily selects a leaf, and calls out "three"; and (4) continues with the same pattern, selecting two (or three) leaves per plant and consecutively counting toward 100.

   e. While the leaf counter calls out numbers, the observer compares each number with the six random numbers on the list. When a leaf number corresponds with a random number, the observer calls out "collect that leaf."

   f. The leaf counter removes each leaf so identified and puts it into a collecting bag. If the plant has compound leaves (e.g., beans), the entire leaf should be collected, not just single leaflets. After a leaf is collected, the leaf counter continues counting where he or she left off. The process continues in a row until six leaves have been collected from the row. If the leaf counter comes to the end of the row before six leaves are collected, he or she should start over at the beginning of the row.

   g. Repeat steps 4b–4f for the other three marked rows. At the end of the collecting process, there should be 24 leaves in the collecting bag.

5. Measure and record herbivore consumption on the 24 sample leaves. The following steps describe how to do this for one leaf.

    Note: If a leaf area meter is not available, herbivore consumption can be measured by following this alternative procedure: Put each leaf on a photocopying machine, photocopy it, cut out the leaf image (the remaining part of the leaf), and weigh it with an analytical balance, and then cut out the areas of the photocopy that correspond to the consumed areas of the leaf and weigh these cutouts as well. The mass of each cutout is assumed to be proportional to the area of each cutout, so these mass measurements can be used in the datasheet instead of area measurements. Since percent consumption is a ratio without units, it does not matter that the units of measurement are grams.

    a.   Cut out a piece of clear acetate slightly larger than the leaf and place it over the leaf.

    b.   Note the area of the leaf that is missing, including parts with only a transparent layer of leaf tissue remaining. With a fine-tip black marker, make an image of this area by coloring it in as precisely and completely as possible. On leaves with extensive damage, this may involve estimating the original outline of the leaf, based on the size and shape of similar unconsumed leaves.

    c.   Feed the colored acetate into the leaf area meter. The readout gives the area of the black region in square centimeters to the nearest tenth of a square centimeter. Record this area in the "Consumed area" (C) column of a copy of the Herbivory Comparison Datasheet (Figure 15.1).

    d.   Feed the entire leaf into the leaf area meter. It measures the entire remaining area of the leaf. Record the measurement in the "Remaining area" (R) column of the datasheet.

    e.   Calculate the percentage of the leaf's area that has been consumed. Add the figures for R and C and record the result in the "Original leaf area" column. Then divide the consumed area (C) by the original leaf area ($C + R$), multiply by 100, and record this figure as a percentage in the rightmost column of the datasheet.

Data Analysis

With each group sharing its data, construct bar graphs that allow the comparison of percent consumption (1) between plots within blocks and (2) among same-type plots between blocks. Find the mean percent consumption among all the monoculture plots and the mean percent consumption among all the intercrop plots and compare these two figures.

    Optional: Determine if the difference in herbivore consumption between the two types of plots within each block is statistically significant using a Student's *t*-test.

Write-Up

The results of this investigation can be reported independently or as a component of the report for Investigation 19.

    In either case, write-ups should incorporate the data collected by all four teams. The following are some suggestions for writing the reports:

- Summarize the results, as described in Data Analysis earlier.
- Discuss what kind of relationship between system diversity and herbivore consumption is indicated by the results.
- Discuss the observed variation in relative herbivore consumption among blocks.
- Note any observations about the consumption patterns of particular insects. (Do some herbivores eat only particular plant species or plant parts?)
- Discuss how the results might be applied to farming.
- Suggest additional related questions to explore and investigate.

	Herbivory Comparison Datasheet			

Sampling Date: October 17, 2015   Crop: Bush Beans

Treatment: Monoculture

Leaf	Consumed Area, cm² (C)	Remaining Area, cm² (R)	Original Leaf Area, cm² (C + R)	% of Leaf Consumed $\left(\dfrac{C}{C + R} \times 100 \right)$
1	0.5	7.5	8.0	6
2	5.3	2.3	7.6	70
3	0	7.4	7.4	0
4	0.8	6.6	7.4	11
5	0.2	7.9	8.1	2
6	0.2	7.6	7.8	3
7	0	6.9	6.9	0
8	0	7.0	7.0	0
9	0.9	7.1	8.0	11
10	0.5	7.7	8.2	6
11	0.3	7.4	7.7	4
12	0	7.7	7.7	0
13	0.9	6.4	7.3	12
14	0.7	6.6	7.3	10
15	0.1	7.7	7.8	1
16	0.3	7.5	7.8	4
17	0.6	7.2	7.8	8
18	0.1	7.8	7.9	1
19	0.5	7.3	7.8	6
20	0.4	7.7	8.1	5
21	0	7.4	7.4	0
22	0.2	7.6	7.8	3
23	0.3	7.1	7.4	4
24	0.7	7.0	7.7	9
			Total	176
Mean percentage of leaf area consumed (total/24)				7%

Figure 15.1
Example of a completed Herbivory Comparison Datasheet.

Variations and Further Study

1. Perform this investigation independently of Investigation 19 by planting two plots that differ in structure and diversity and by comparing the herbivore consumption in each at various points in their development. The following table suggests some useful comparisons. In each case, the two compared plots should be similar in all other respects. The best results will be obtained with fairly large plots (i.e., larger than 0.2 ha, or about 40 × 50 m) with 10 m wide crop-free strips between plots.

Less Diverse System	More Diverse System
Weeded plot	Unweeded plot
Monoculture	Polyculture
Portion of a much larger monoculture	Monoculture plot within a highly diverse mosaic of different monocultures
Plot lacking interspersed rows	Plot with interspersed rows of perennials, trees, or tall annuals

2. Find two existing systems (other than the Investigation 19 intercrop plots) on which this investigation could be performed. Any two systems containing the same crop but differing in structure or diversity could be used.

3. As an additional procedure, determine the total amount of leaf biomass removed from each treatment. The first step in this process is to dry the sampled leaves from each plot (48 h at 60°C) and find the combined mass of each plot's sample. Once this remaining mass is determined, the following equation can be used to find the original mass:

$$\text{Original mass} = \frac{\text{Remaining mass}}{1 - \% \text{ consumption}}$$

The remaining mass is then subtracted from the original mass to find the mass of leaf matter consumed:

$$\text{Original mass} - \text{remaining mass} = \text{Consumed mass}$$

Once the consumed mass is known, the average consumed mass per leaf can be easily calculated:

$$\text{Average consumed mass per leaf} = \frac{\text{Consumed mass of sample}}{24}$$

Then all that remains is to estimate the number of leaves in the plot, which involves sampling a number of plants to determine the average number of leaves per plant and multiplying this average by the estimated or actual number of plants in the plot. The number of leaves in the plot is then multiplied by the average consumed mass per leaf to determine the total leaf biomass consumed:

$$\text{Average consumed mass per leaf} \times \text{number of leaves in plot} = \text{total leaf biomass consumed}$$

4. Measure leaf damage as well as herbivory during the data collection phase of the investigation. Damaged leaf area is that which still exists (and is detected by the leaf area meter) but is no longer functioning; it includes diseased tissue, tissue partially eaten by leaf-miner-type insects, and necrotic tissue at the edge of areas eaten by herbivores. Some leaves may have a significant amount of damaged tissue.

Leaf damage should be measured separately from herbivory. The desired figure for each leaf is the percent of the *remaining* leaf area (R) that is damaged. To derive this result, (1) trace the damaged tissue of each leaf on clear plastic and measure its area as D; (2) divide D by R, as measured for the calculation of percent consumption; and (3) multiply the result by 100.

$$\% \text{ Damage to the remaining leaf} = \frac{\text{Area of leaf damage}}{\text{Remaining leaf area}} \times 100$$

Herbivory Comparison Datasheet

Sampling Date: Crop:

Treatment:

Leaf	Consumed Area, cm² (C)	Remaining Area, cm² (R)	Original Leaf Area, cm² (C + R)	% of Leaf Consumed $\left(\dfrac{C}{C+R} \times 100 \right)$
1				
2				
3				
4				
5				
6				
7				
8				
9				
10				
11				
12				
13				
14				
15				
16				
17				
18				
19				
20				
21				
22				
23				
24				
			Total	
Mean percentage of leaf area consumed (total/24)				

Investigation 16

Herbivore Feeding Preferences

Background

When an herbivorous insect encounters a monoculture of a crop for which it has a food preference, that insect can rapidly increase its population size, inflict considerable damage to the crop, and become a serious pest. Conventional farmers have come to rely greatly on synthetic chemical pesticides to control such outbreaks, but the negative impacts of the use of these inputs have become well known. They include the development of resistance by the target insects and negative effects on nontarget species such as other insects, nonagricultural species in the environment, human consumers of agricultural products, and the farmers and farmworkers who grow the food for us (Chapter 1 of *Agroecology*: *The Ecology of Sustainable Food Systems*).

Agroecological alternatives to synthetic pesticides, designed to lower the numbers of pest insects while lessening their impacts on the crop, depend on a range of beneficial interactions between different components of the agroecosystem. These interactions take many forms (Chapter 16 of *Agroecology*: *The Ecology of Sustainable Food Systems*), but only come about when diversity and complexity are a central priority in agroecosystem management. Only with high diversity is there a potential for beneficial interactions.

Many organic farmers seek solutions to their pest problems by merely substituting a synthetic chemical with a natural pesticide that is acceptable to organic production standards. But the conditions that attract pests to their fields in the first place continue to exist, and pest management remains a major problem. However, many methods of "alternative" pest management that provide excellent examples of how establishing a foundation of diversity and complexity can lead to beneficial interactions that keep pest populations in check have been developed (Chapter 17 of *Agroecology*: *The Ecology of Sustainable Food Systems*).

A key part of any pest management strategy, especially one that incorporates high diversity, is learning more about the life history and behavior of each pest. With knowledge of how a pest responds to an array of food options, for example, a farmer can design an agroecosystem to include species that draw the pest away from the primary crop plant of the system.

Textbook Correlation

Investigation 14: The Population Ecology of Agroecosystems (Applications of Niche Theory to Agriculture: Biological Control of Insect Pests)

Investigation 16: Species Interactions in Crop Communities

Investigation 17: Agroecosystem Diversity

Synopsis

A locally abundant generalist insect herbivore is placed in petri dishes containing varying sets of food options: either one preferred species, a mixture of the preferred species and another crop plant, or a mixture of the preferred species and a noncrop plant. Feeding activity by the insects is measured by the amount of leaf surface area consumed or damaged. Consumption of the preferred crop plant is compared across the range of plant mixtures, and the results used for understanding the possible role of diversity in pest management.

Objectives

- Gain experience manipulating an insect herbivore under controlled experimental conditions.
- Investigate herbivore feeding preferences in simple and complex systems.
- Correlate laboratory findings to possible conditions in the field.

Procedure Summary and Timeline

Prior to day 1

- Locate an abundant source of a local pest herbivore, a range of plant food sources for this pest, and other materials for the investigation.

Day 1

- Cut leaf discs and set up feeding preference chambers.

Day 2

- Compare feeding rates in each chamber.

After data collection

- Analyze data and write up results.

Note: Unlike most other investigations, data are collected only 24 h after setup. Students must be able to be present in the lab on two successive days.

Timing Factors

Although this investigation is carried out in the lab, it depends on the availability of a generalist herbivore and crop plants as food options; hence, it may be restricted to a time of the year when cropping is occurring. It lends itself to rapid setup, observation, and takedown.

Materials, Equipment, and Facilities

Sixty 9 cm diameter petri dishes (glass or plastic)

120–200 individuals of an herbivorous insect species

Fresh leaves of four crop species known to be eaten by the insect

Fresh leaves of another crop species

Fresh leaves of one noncrop (weed) species

Cork borer, 1.0–1.5 cm diameter

Several large rubber corks

Sixty small pieces of sponge

Sheets of graph paper or a leaf area meter

Three or four gallon jars with lids

Marking tape

Marking pens

Advance Preparation

- Obtain the nonliving materials.
- Identify an herbivorous insect species appropriate for the investigation. The insects that will work well for this investigation are the generalist herbivores, such as the spotted or striped cucumber beetle (*Diabrotica* spp.), grasshoppers, and larvae of several butterflies or moths, such as the cabbage looper (*Trichoplusia ni*) or the diamondback moth (*Plutella* sp.). Two or three individuals will be needed for each petri dish (120 minimum total), so make sure there is an abundant local source.
- In the early morning of the day of the setup, before they have had an opportunity to do much feeding, collect 120–200 individuals of the identified insect. It may be even better to collect the insects 2 or 3 days before the setup of the investigation, keeping them in large jars with bits of moist sponge, so that they will be hungrier on the day of the investigation setup.
- On the day of the investigation setup, collect several whole fresh leaves of each of the following: four crops known to be eaten by the pest, a crop that could be intercropped with the preferred crop, and a noncrop or weed species. The potential intercrop species can be one known to be eaten by the pest, but it is best to use one that is not normally fed upon by the pest. The noncrop or weed species can be one that might be fed on by the pest or one that might repel the pest. All the leaves must be at least as wide as the diameter of the cork borer (1–1.5 cm), which excludes some narrow-leaved and compound-leaved species.
- Cut sponges into squares approximately 2 cm × 2 cm, with a thickness of less than 1 cm so they will fit inside the petri dishes.

Ongoing Maintenance

None required.

Investigation Teams

Form four teams, each made up of three or four students. Each team will be responsible for testing the feeding behavior of the insect using a different preferred crop species, along with the same intercrop and noncrop species used by the other teams. See the following logistics map.

Team 1	Team 2	Team 3	Team 4
Preferred crop A	Preferred crop B	Preferred crop C	Preferred crop D
3 treatments × 5 replicates = 15 test chambers	3 treatments × 5 replicates = 15 test chambers	3 treatments × 5 replicates = 15 test chambers	3 treatments × 5 replicates = 15 test chambers

Procedure

The successful operation of this investigation depends on several variables, and being aware of the possibility of unusual or mixed results is critical to its success.

The following steps describe the setup and data collection procedures for one team's set of three treatments.

Setup

1. Set out 15 petri dishes in three rows of five dishes each. Place a small piece of marking tape on the lid of each dish, and label each one with a treatment type and replicate number as follows:

   a. Crop only, replicates 1–5

   b. Crop/intercrop mixture, replicates 1–5

   c. Crop/weed mixture, replicates 1–5

2. Obtain several leaves of the preferred crop species assigned to the team and a few leaves each of the intercrop species and the weed species.

3. Place a leaf of the preferred crop species on the top of a large rubber cork, and with a sharp cork borer (diameter between 1 and 1.5 cm), cut out a leaf disc. Using this technique, make a total of 40 discs of the preferred crop, 10 discs of the intercrop species, and 10 discs of the weed species. If the types of discs cannot be easily distinguished, devise a system of small ink markings to tell them apart.

4. Set up the test chambers:

   a. Place a piece of moistened sponge in the center of the dish labeled "Crop only, replicate 1."

   b. Place four preferred-crop leaf discs in the bottom of the dish, equally spaced around the sponge and about 1 cm from the edge of the dish. Make sure no disc touches the sponge.

   c. Repeat steps a and b for the other four replicates of the crop only treatment.

   d. Set up the five replicates of the crop/intercrop mixture treatment in the same way as those of the crop only treatment, except substitute two intercrop leaf discs for two of the preferred-crop discs. The different types of discs should alternate around the dish.

   e. Set up the five replicates of the crop/weed mixture treatment in the same way, placing two preferred-crop leaf discs and two weed leaf discs in each dish.

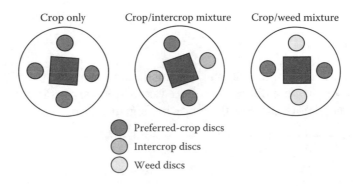

Figure 16.1
Setup of the three treatments.

5. Release three of the insects into each dish, being careful to replace the lid quickly if dealing with insects capable of flight. If there are not enough insects available for three per dish, put two into each dish (but do not put three in some dishes and two in others).

6. Put the dishes in a well-lit space in the laboratory, away from direct sunlight, and where they will be kept at room temperature for 24 h.

Data Collection

The following steps are performed 24 h after setting up the feeding chambers:

1. One at a time, open a dish, remove the leaf discs, and, using one of the methods described in the following, measure the consumed area of each disc. (Note that some insects may eat only the surface cells and not leave holes. In this case, use the method described in Investigation 15 to determine the area of consumption.)

Method A: Leaf Area Meter

   a. Feed a freshly cut, undamaged disc into the meter. Record the readout, which will give the area of the disc in square centimeters.

   b. Feed a partially consumed disc into the leaf area meter and record the readout.

   c. Subtract the area of the partially consumed disc from the previously recorded area of a whole disc to get the area of the disc (in square centimeters) that has been consumed.

   d. Convert the square-centimeter area to square millimeters by moving the decimal two places to the right.

   e. Repeat steps b through d for the other partially consumed discs in the dish.

Method B: Graph Paper

   a. Place a partially consumed disc onto a sheet of graph paper.

   b. Outline the area of the disc that has been consumed.

   c. Count the number of whole squares in the traced area, estimate the number of squares taken up by the partial squares, and add the two numbers.

   d. If not using 1 mm graph paper, convert the total number of squares into a square-millimeter measurement.

   e. Repeat steps a through d for the other partially consumed discs in the dish.

2. Record each measurement of area consumed on the Feeding Preference Datasheet (Figure 16.2).

Data Analysis

1. Calculate the total disc area consumed in each dish.

2. Calculate the mean leaf disc area consumed per disc in each of the dishes of the crop only treatment.

3. Calculate the mean leaf disc area consumed per disc for the two crop discs in each of the dishes of the crop/ intercrop treatment and the crop/weed treatment. (Note that each mean will be found by dividing the total consumption of the two crop discs by 2; do not use the total calculated in step 1.)

4. Calculate the mean leaf disc area consumed per disc for the two intercrop discs in each of the dishes of the crop/ intercrop treatment.

5. Calculate the mean leaf disc area consumed per disc for the two weed discs in each of the dishes of the crop/ weed treatment.

6. Calculate the treatment means for each measure of consumption (i.e., each row of the datasheet table).

7. Compare the consumption of the crop species discs across the three treatments.

8. Share your team's data with the other teams, and acquire other teams' data for use in writing the report.

	Feeding Preference Datasheet						
Preferred Crop: Broccoli			Herbivore: Cabbage Looper				
Weed Species: Wild Mustard			Intercrop Species: Lettuce				

Treatment	Discs	Replications					Treatment Means
		1	**2**	**3**	**4**	**5**	
Crop only	Crop disc 1 consumption (mm^2)	123	89	145	67	98	104.4
	Crop disc 2 consumption (mm^2)	111	78	140	103	110	108.4
	Crop disc 3 consumption (mm^2)	124	94	117	87	130	110.4
	Crop disc 4 consumption (mm^2)	76	98	120	111	121	105.2
	Total consumption (mm^2)	434	359	522	368	459	428.4
	Mean area consumed per disc (mm^2)	108.5	89.75	130.5	92.0	114.75	107.1
Crop/intercrop	Crop disc 1 consumption (mm^2)	101	89	65	79	92	85.2
	Crop disc 2 consumption (mm^2)	97	76	98	45	52	73.6
	Intercrop disc 1 consumption (mm^2)	23	10	31	25	27	23.2
	Intercrop disc 2 consumption (mm^2)	16	19	31	26	20	22.4
	Total consumption (mm^2)	237	194	225	175	191	204.4
	Mean area consumed per disc, crop discs (mm^2)	99.0	82.5	81.5	62.0	72.0	79.4
	Mean area consumed per disc, intercrop discs (mm^2)	19.5	14.5	31.0	25.5	23.5	22.8
Crop/weed	Crop disc 1 consumption (mm^2)	65	49	30	75	52	54.2
	Crop disc 2 consumption (mm^2)	73	51	29	53	71	55.4
	Weed disc 1 consumption (mm^2)	150	123	146	180	162	152.2
	Weed disc 2 consumption (mm^2)	161	176	119	167	156	155.8
	Total consumption (mm^2)	449	399	324	475	441	417.6
	Mean area consumed per disc, crop discs (mm^2)	69.0	50.0	29.5	64.0	61.5	54.8
	Mean area consumed per disc, weed discs (mm^2)	155.5	149.5	132.5	173.5	159.0	154.0

Figure 16.2
Example of a completed Feeding Preference Datasheet.

Write-Up

The following are some suggestions for reporting on the results of the investigation:

- Present the results for each preferred-crop test in a way that allows comparison of consumption across the three treatments.
- Discuss the impact of diversification on feeding activities of pests.
- Discuss the implications of the results for the design of alternative pest management strategies.

Variations and Further Study

1. Carry out the investigation using different species of herbivores.
2. Design a similar but longer-term investigation using potted plants kept in lighted, screened boxes into which herbivorous insects are released. Put only one species of crop plant into some boxes and mixtures of plants into others. The additional plants could be repellent plants, noncrop trap plants, and other crop plants.
3. Carry out a similar study as part of Investigation 19. Compare damage on the leaves of each crop in monoculture to damage on the leaves of the same plant in the intercrop plots.
4. Design a way to test how a noncrop plant that is known to be attractive to herbivores could be used effectively as a decoy or trap crop, taking pressure off of the main crop when in association with it.

Feeding Preference Datasheet							
Preferred Crop:				Herbivore:			
Weed Species:				Intercrop species:			

Treatment	Discs	Replications					Treatment Means
		1	2	3	4	5	
Crop only	Crop disc 1 consumption (mm^2)						
	Crop disc 2 consumption (mm^2)						
	Crop disc 3 consumption (mm^2)						
	Crop disc 4 consumption (mm^2)						
	Total consumption (mm^2)						
	Mean area consumed per disc (mm^2)						
Crop/intercrop	Crop disc 1 consumption (mm^2)						
	Crop disc 2 consumption (mm^2)						
	Intercrop disc 1 consumption (mm^2)						
	Intercrop disc 2 consumption (mm^2)						
	Total consumption (mm^2)						
	Mean area consumed per disc, crop discs (mm^2)						
	Mean area consumed per disc, intercrop discs (mm^2)						
Crop/weed	Crop disc 1 consumption (mm^2)						
	Crop disc 2 consumption (mm^2)						
	Weed disc 1 consumption (mm^2)						
	Weed disc 2 consumption (mm^2)						
	Total consumption (mm^2)						
	Mean area consumed per disc, crop discs (mm^2)						
	Mean area consumed per disc, weed discs (mm^2)						

17

Effects of a Weedy Border on Insect Populations

Background

A body of literature has been accumulating in recent years supporting the belief that introducing specific forms of diversification into an agroecosystem can have many beneficial effects (Chapter 17 of *Agroecology: The Ecology of Sustainable Food Systems*). This literature contradicts a basic principle of modern industrial agriculture: to simplify agroecosystems by eliminating both noncrop plants (usually referred to as weeds) and any insects.

One important form of diversification is the deliberate inclusion of certain weeds in the agroecosystem (Chapter 16 of *Agroecology: The Ecology of Sustainable Food Systems*). The strategic placement of weeds in an agroecosystem can affect the ability of pests to find and colonize a crop, or it can attract beneficial insects that can help control pests. When weeds are used as borders around a crop, the concepts of island biogeography (Chapter 17 of *Agroecology: The Ecology of Sustainable Food Systems*) can come into play. The weed may act as a dispersional barrier around the crop "island." As a barrier, the weedy border may "trap" pests or shield the crop from detection and colonization by pests. Or the weed border may itself act as an island of suitable habitat for beneficial insects, which may disperse outward into the crop to prey on pests.

Textbook Correlation

Investigation 14: The Population Ecology of Agroecosystems
Investigation 16: Species Interactions in Crop Communities
Investigation 17: Agroecosystem Diversity

Synopsis

Seeds of a weed species are planted as a border along a crop at the time that the crop is planted out to the field. After a period of development of both the crop and the weed border, insects are censused, and damage to the crop is assessed at different distances from the weedy border. The diversity of insects, both pest and beneficial, is compared to those in a crop planted without the border.

Objectives

- Learn insect sampling techniques.
- Learn methods for assessing herbivore damage on a crop.

- Learn how to look for signs of predation and/or parasitization of a crop pest in the field.
- Compare insect populations in crops with and without weedy borders.
- Determine the distance into a crop that a strip of border vegetation has an effect.
- Gain an understanding of the role of island biogeography theory in the design and management of farming systems.

Procedure Summary and Timeline

Prior to week 1
- Obtain weed seeds; obtain seed of crop; prepare farm field for planting; sow crop seed in speedling trays; sow weed seed in field border; transplant crop seedlings into field.

Week 1
- Make observations of insect presence and crop damage.

Week 4
- Make observations of insect presence and crop damage.

Week 7
- Make final observations of insect presence and crop damage.

After data collection
- Analyze data and write up report.

Timing Factors

This investigation must be carried out during the growing season, is set up prior to the beginning of the period of instruction, and requires 11–13 weeks from initial setup to completion. In addition, the crops and weeds selected may perform best during a particular part of the growing season.

Materials, Equipment, and Facilities

Field space, approximately 25 m × 65 m

Machinery for cultivating soil and forming bedded-up rows

About 4000 seeds (or 3200 + transplants) of cauliflower

Seeds of corn spurry, *Spergula arvensis*

Lathhouse or greenhouse for starting seedlings

Twenty speedling trays (to accommodate approximately 4000 seedlings)

About 80 lb of soil mix suitable for starting seeds in speedling trays

Hand-operated mechanical seed planter

Metric tape measure

Stakes and string for delineating section borders

Butterfly nets

Insect guides for identification

Three copies of the Weed Border Effects Datasheet, per student or team

Advance Preparation

The seeds for this investigation must be sown 5–6 weeks before the first intended data collection date, so plan accordingly. Note also that the corn spurry seed may have to be collected a year ahead of time.

- Choose a weed species and a crop species for this investigation. The investigation is written to employ corn spurry (*S. arvensis*) as the weed and cauliflower as the crop, because the author has experience with these species on the central coast of California and knows they produce good results. However, instructors in other regions and climates are encouraged to select alternate species more appropriate to the local conditions and the time of year when the investigation is carried out. Choosing appropriate species will require some knowledge of the relationships among crops, weeds, pests, and beneficials in the local area. Instructors using species other than cauliflower and corn spurry should ignore subsequent references to these specific species and their associated pests and beneficials.

- Obtain seed of corn spurry (*S. arvensis*) and a variety of cauliflower, such as "snowball." If the corn spurry seed cannot be obtained commercially, it should be collected the previous season by pulling mature plants, letting them dry in bags, and then shaking the seeds out.

- Secure field space approximately 25 m × 65 m. The entire space should be uniform, with no differences in the surrounding vegetation.

- Counting back 5 or 6 weeks from the week during which you want data collected for the first time, choose a date on which to sow the cauliflower seeds in trays and the weed seeds in the field.

- Mark the boundaries of the study plots in the field. Figure 17.1 shows recommended dimensions and layout.

- Prepare the field areas for planting. Amend the soil as necessary and cultivate. In the area in which cauliflower is to be transplanted, form lengthwise bedded-up rows with crowns approximately 0.8 m apart. In the 1 m strip in which corn spurry is to be sown, simply cultivate and level the soil. Do the same for a 5 m barrier area between the two plots.

- On the previously planned sowing date, sow the corn spurry seeds in the weed border planting area using the hand broadcast method. Try to keep the seed density fairly constant.

- On (or near) the same day as the corn spurry seeds are sown in the field, sow the cauliflower seeds in speedling trays. Approximately 3400 seedlings will be needed, so sow the number of seeds needed to yield at least that many seedlings. Keep the trays in a lathhouse or similar area, where conditions will be suitable for germination and seedling growth.

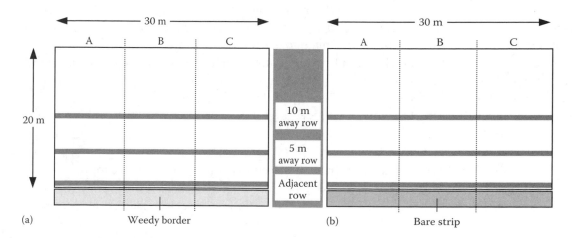

Figure 17.1
Layout of the experimental (a) and control plots (b) showing sections and transects.

- Water both the corn spurry bed in the field and the cauliflower trays.
- When the cauliflower seedlings are approximately 10 cm tall, transplant them out into the plots. Space the seedlings approximately 50 cm apart along the previously formed rows. Make sure that both plots are planted identically.
- Reserve several dozen cauliflower seedlings for replacing plants that die soon after plant-out.
- Use stakes and tape or string to delineate the three sections (A, B, and C) of each plot, as shown in Figure 17.1. Assign a section to each team.
- Make the appropriate number of copies of the datasheet.

Ongoing Maintenance

The cauliflower plots should be kept free of weeds for the duration of the investigation, and other weeds besides corn spurry should be kept out of the weed border. Keep the bare strip along the control plot bare. Irrigate the plots and the weed border as necessary for local conditions. Control noninsect pests (e.g., gophers) if they are a local problem. Do not apply any pesticides near the plots during the investigation—the plots should be at least 100 ft from any sprayed fields.

Investigation Teams

Form three teams, each with four to eight members. Each team will be responsible for collecting data in one section of the weed-bordered (experimental) plot and the corresponding section of the unbordered (control) plot.

Procedure

Data Collection

At weeks 1, 4, and 7 (or at similar intervals), perform each of the observations/measurements described in the following in your team's section of both the weed-bordered plot and the unbordered control plot. In the weed-bordered plot, do so (1) in the row of cauliflower adjacent to the weed border, (2) in the row of cauliflower approximately 5 m from the border, and (3) in the row approximately 10 m from the border. In the control plot, perform the measurements in the analogous rows of the crop.

For each type of measurement, the sampling procedure is left somewhat open. It is up to the team to devise an appropriate way of doing representative sampling in each row.

Use a different copy of the datasheet for each day of data collection.

The following steps describe the sampling procedure for one row/transect. They will be repeated for the other two rows/transects.

1. Count the number of aphid colonies on 10 randomly selected cauliflower plants in the row.
   a. On each plant selected, count the number of aphid individuals and record this figure in your lab notebook. If infestation is especially high, it may be easiest to count the number of individuals in a representative colony and then multiply this number by the number of colonies.
   b. Calculate the average number of aphids per plant and record this figure on a copy of the Weed Border Effects Datasheet (Figure 17.2).

Weed Border Effects Datasheet

Sampling Date: June 30, 2015

Weedy Border Plot

	Adjacent to Weed Border				5 m from Border				10 m from Border			
	Sections				Sections				Sections			
	A	B	C	Mean	A	B	C	Mean	A	B	C	Mean
Aphids per plant	1.6	1.4	1.5	1.5	5.1	4.4	4.7	4.7	10.2	9.5	10.1	9.9
Ratio of live to parasitized aphids	6:1	7:1	7:1	7:1	9:1	10:1	9:1	9:1	10:1	11:1	9:1	10:1
% damage per leaf	3.4	3.9	2.9	3.4	6.2	6.7	6.4	6.4	8.6	9.9	9.2	9.2
Larvae per plant	0.9	0.2	0.6	0.6	2.4	2.5	2.2	2.4	3.2	2.9	3.0	3.0
Eggs per plant	4.0	3.8	3.7	3.8	5.0	3.0	4.3	4.1	5.5	5.6	5.1	5.4
Pupae per plant	0.5	0.4	0.2	0.4	0.8	1.0	0.5	0.8	1.6	1.5	1.8	1.6

Control plot

	Adjacent to Weed Border				5 m from Border				10 m from Border			
	Sections				Sections				Sections			
	A	B	C	Mean	A	B	C	Mean	A	B	C	Mean
Aphids per plant	7.3	8.4	7.8	7.8	10.2	11.4	9.9	10.5	10.5	11.4	9.6	10.5
Ratio of live to parasitized aphids	20:1	22:1	18:1	20:1	20:1	21:1	19:1	20:1	23:1	18:1	19:1	20:1
% damage per leaf	9.4	9.7	9.1	9.4	10.0	9.4	9.9	9.8	10.4	10.3	11.1	10.6
Larvae per plant	2.7	3.1	2.6	2.8	3.1	3.3	2.8	3.1	3.0	2.9	3.1	3.0
Eggs per plant	4.3	5.0	3.1	4.1	5.2	5.5	4.2	5.0	5.4	6.1	4.5	5.3
Pupae per plant	0.5	0.8	0.4	0.6	1.4	1.2	1.1	1.2	1.5	1.4	1.2	1.4

Figure 17.2
Example of a Completed Weed Border Effects Datasheet.

2. Determine the proportion of parasitized aphids to live aphids on 10 randomly selected leaves in the row:
   a. On each leaf selected, count the number of parasitized aphids, which appear as dried, hardened, thickened, gray or light brown "mummies." Then count or estimate the total number of live aphids on the leaf. Record the two observations as a ratio in your lab notebook.
   b. Calculate the average ratio of live to parasitized aphids on the 10 leaves examined and record this ratio on the datasheet.

3. Measure the amount of leaf damage on 10 randomly selected leaves in the row.
   a. The amount of damage per leaf can be estimated visually (as a rough %), or each leaf can be collected, and the methods described in Investigation 15 can be used to measure leaf damage more precisely.
   b. Calculate the mean percentage of damage per leaf and record this figure in the datasheet.

4. Count the number of lepidopteran larvae (caterpillars) on 10 randomly selected plants in the row.
   a. Examine every plant surface—stems and upper and lower surfaces of leaves—for the presence of larvae, which may be the same green color as the leaves. Record the number of larvae per plant in your lab notebook. (You may want to distinguish between species, which may include the imported cabbage worm, *Pieris rapae*; the cabbage looper, *Trichoplusia ni*; and the diamondback moth, *Plutella maculipennis*.)
   b. Calculate the mean number of larvae per plant and record this figure in the datasheet. If you have divided your count by species in your lab notebook, report a combined number on the datasheet.

5. Count the number of lepidopteran eggs on 10 randomly selected plants in the row.

     a. Imported cabbage worm eggs are usually laid on the undersides of leaves and appear thin, yellow, and about 2 mm long. The eggs of the cabbage looper are flattened, round, translucent, and about 1 mm in diameter. The eggs of the diamondback moth are very small and usually laid in clusters of 10–15. Inspect the underside of every leaf on each plant, and record the number of eggs on each plant in your lab notebook. Divide your count by species if desired.

     b. Calculate the mean number of eggs (of all species) per plant and record this figure on the datasheet.

6. Count the number of pupae (cocoons or chrysalises) of the cabbage worm, cabbage looper, and diamondback moth on 10 randomly selected plants in the row.

     a. The pupae are often found on the undersides of leaves. Count the number of pupae on each plant and record the number in your lab notebook. If you can distinguish species, divide the data by species.

     b. Calculate the mean number of pupae (of all species) per plant and record this figure in the datasheet.

     Perform the following procedures during each sampling period in the weedy border adjacent to your team's section of the weedy border plot.

7. Observe the developmental stage of the plants in the weedy border. Take notes on plant height and number of open flowers.

8. If any of the *Spergula* plants have open flowers, observe the numbers and types of insects visiting the flowers in a 1 m² area during a 30 min period. These observations should be made at the same time each day. Late morning may be best, because this is when flowers are open and actively producing nectar and pollen.

     a. Choose a 1 m² section of the weed border randomly and mark it off with string and stakes.

     b. Carefully watch for insect activity in the quadrat. If you cannot identify the types of insects visiting the flowers, collect several with a butterfly net for later identification in the lab.

     c. Record the data in your lab notebook for use in writing up the results of the investigation.

Data Analysis

1. Share your data with the other teams, and use other teams' data to fill in the blank cells in the datasheets for each sampling date.

2. Calculate the mean of each measurement across the three sections of each plot for each sampling date and record the means on the datasheets.

3. *Optional*: Perform statistical analyses of the variance among the measurements from each section of each plot.

Write-Up

The following are some suggestions for reporting on the results of the investigation:

- Present data in graph form to show how insect numbers and damage varied according to distance from the weed border in both the experimental and control plots.
- Explain spatial and temporal patterns in pest distribution.
- Evaluate whether the data indicate an effect from the weed border.
- Discuss differences between sections within each plot and what factors may account for the differences.
- Discuss how the results could be applied in agriculture.

Variations and Further Study

1. Perform the investigation with a weed border made up of weeds that come up spontaneously *in situ*.

2. Sample every 5 m from the weed border to the opposite edge of the plot. Smaller distances between sample points may give an even more detailed idea of the distribution of insects, as impacted by the border.

3. Use a wider range of insect sampling methods to get a better idea of how the total insect community is affected by the weed border. (See Investigation 11 for descriptions of other sampling methods.)

4. Plant three or four different species of weeds as borders along the crop, and compare the effects on both pests and beneficials for each one.

5. Try borders of differing widths.

6. Try sowing the weed seed at different times in relation to when the crops is sown or transplanted out; for example, 2 weeks before transplanting, at plant-out, and 2 weeks after.

7. Rather than planting the weed in a bordering strip, disperse the weed among the crops at different densities to determine an optimum density that will not interfere with the crop, but will still affect the insects.

8. Collect parasitized insects (aphid mummies or lepidoptera larvae) and bring them to the laboratory to observe for later emergence and identification of the parasitic organism.

Weed Border Effects Datasheet

Sampling Date:

Weedy Border Plot

	Adjacent to Weed Border				5 m from Border				10 m from Border			
	Sections				Sections				Sections			
	A	B	C	Mean	A	B	C	Mean	A	B	C	Mean
Aphids per plant												
Ratio of live to parasitized aphids												
% damage per leaf												
Larvae per plant												
Eggs per plant												
Pupae per plant												

Control plot

	Adjacent to Weed Border				5 m from Border				10 m from Border			
	Sections				Sections				Sections			
	A	B	C	Mean	A	B	C	Mean	A	B	C	Mean
Aphids per plant												
Ratio of live to parasitized aphids												
% damage per leaf												
Larvae per plant												
Eggs per plant												
Pupae per plant												

Studies of Farm and Field Systems

18

Mapping Agroecosystem Biodiversity

Background

Each agroecosystem has its own unique structure and spatial organization. In a diverse system, the spatial organization of just the biotic components of the system may be exceedingly complex, and understanding it requires that one ask a variety of questions: What species are present? What is each species' relative abundance? How are the species distributed over space? What is the role of each species in the ecology of the system? What patterns are created by the distribution of species? How do these patterns change over time?

The answers to such questions are the basis for a kind of biogeography of the agroecosystem. Biodiversity maps of diverse agroecosystems are important because they help us understand how diversity is actually manifested and how it might be translated into sustainable function.

Textbook Correlation

Investigation 17: Agroecosystem Diversity
Investigation 21: Landscape Diversity

Synopsis

An agroecosystem is examined, surveyed, and mapped. The map is used as a basis for recognizing patterns of species distribution and abundance, interspecific relationships, and diversity. It can later be used as a basis for studying ecological relationships, ethnobotany, and farmer decision making about agroecosystem design and management.

Objectives

- Practice observational skills and improve the ability to recognize and describe spatial patterns.
- Learn basic techniques of mapping an agroecosystem.
- Learn the taxonomy of species, both crop and noncrop, in an agroecosystem.
- Analyze patterns of species abundance, species distribution, patchiness, and spatial variation.
- Analyze patterns of interspecific relationships.

Procedure Summary and Timeline

Prior to week 1

- Select the agroecosystem(s) to be mapped.

Week 1

- Survey and analyze the system, set up transect lines, and collect mapping data.

Week 2

- Complete the aforementioned activities as needed; ask a farmer or gardener for additional information.

After week 2

- Draw maps and write up discussion.

Timing Factors

This investigation can be completed in a relatively short period of time, with little preparation. It can be done anytime during the growing season.

Coordination with Other Investigations

This investigation can be combined with Investigation 23, with the interview process serving as a way of expanding on the knowledge gained by observing and mapping an agroecosystem.

Materials, Equipment, and Facilities

One or more existing agroecosystems suitable for mapping

String or cord for transect lines

50 or 100 m tape measures

Graph paper

Clipboards

Advance Preparation

- Select the agroecosystem(s) to be mapped. Nearly any kind of agroecosystem can be mapped from a relatively small garden to a many-hectare conventional farm. However, the most interesting and instructive systems for this investigation are those with the highest degree of heterogeneity, such as tropical homegarden systems (see Chapter 18 of *Agroecology: The Ecology of Sustainable Food Systems*), diverse small farms, and intensive production systems on the fringes of urban areas. The typical large-scale conventional monoculture system is too simple and homogenous to be a good subject for this investigation.

- If necessary, receive permission from the owner(s) or manager(s) for investigation teams to be on the site(s) for mapping activities.

- If necessary, arrange mutually agreeable times with owners or managers for mapping activities to occur.

Ongoing Maintenance

No maintenance is required.

Investigation Teams

The size and number of teams depend on the system or systems to be mapped. If a single large and complex system is being mapped, each of its major parts can be mapped by a separate team. If several smaller and discrete systems are to be mapped, each can be handled by a separate team. A team can be made up of anywhere from two to eight individuals, depending on the complexity of the system or subsystem to be mapped and the degree of detail and exhaustiveness desired in the analysis.

Procedure

Data Collection

This stage of the investigation involves two major parts—(1) an information-gathering survey of the agroecosystem and (2) a formal mapping of the system. These two components are complementary: the survey informs the mapping process, and the map provides a framework for presenting the results of the survey.

1. Define and delimit the agroecosystem to be mapped and analyzed.
2. Examine the general features and spatial layout of the agroecosystem to become familiar with it.
3. Survey the entire system in some detail, paying attention to its components, their arrangement or structure, and system dynamics. This process may include rough mapping, species identification and cataloging, and obtaining information about plant names and uses from the owner, farmer, farm manager, or other relevant personnel. The following questions serve as a guide for what to be looking for; some may not be relevant for the system being studied.

   a. *System components*: What are the crop species present? How are they divided among annuals and perennials? Among the perennials, what are the growth forms (trees, shrubs, vines, etc.)?

   b. What are the uses of the crop species? Potential usage categories include the following: cash crop, subsistence crop, forage, animal feed, ornamental, home medicinal, and home cooking (e.g., spices and herbs).

   c. What noncrop species are deliberately included in the system? What are the uses of these plants? Potential usage categories include the following: living fencerow, cover crop, firewood, building material, shade, windbreak, and ornamental.

   d. What animals are present in the system? How are they integrated with the cropping components?

   e. *System structure*: How is the system divided into areas with different functions or crop types?

   f. How is each functional area laid out and managed? What are its components? How diverse is it? Is it monocropped or intercropped? What is the spatial arrangement of the plants? Are there different canopy levels?

   g. What is the relationship between the intensely managed parts of the system and their immediate surroundings? Do areas of natural vegetation exist in or around the managed areas? How do they contribute to the diversity of the agroecosystem?

4. Map the system.

    a. Construct a transect grid. A suggested method is to lay parallel transect lines across the width of the system, at 1, 5, 10, or 20 m intervals (depending on the size of the system and the time available). The transect lines should originate from a straight edge that can serve as a baseline reference. One end of the baseline will serve as the 0,0 point of the grid. The transect lines are then marked with tape at similar intervals to create the grid.

    b. Develop a set of symbols for common species, plant types, bed types, trees, fences, paths, and so on.

    c. Decide on a scale for the map (e.g., 1 m on the ground = 1 cm on the map).

    d. Draw a scale representation of the transect grid on a sheet of graph paper.

    e. Locate important individual features such as trees, beds, rows, and borders on the map by reference to their location on the grid. Every location on the ground will be identified by x, y coordinates—for example, 10 m from the baseline and 23 m along the baseline.

    f. If possible, fill in finer detail as possible (e.g., by laying out finer transect lines within complex beds).

    g. Use a compass to orient the transect grid relative to the compass directions.

    h. If the system is on a slope or includes much variation in elevation, use a transit or similar instrument to map the vertical dimension, and show this dimension as one or more cross-sectional diagrams.

Write-Up and Presentation of Data

The results of the data collection, mapping, and system analysis may be presented in the form of a team report. The two major components of a report are an overall map of the system and a discussion of the system's components and structure. An example of an agroecosystem map is shown in Figure 18.1.

- The discussion should describe the system clearly and in some detail and should include a species list, information on the relative or absolute abundance of each species, and descriptions of horizontal and vertical distribution patterns.

- The map should show the major functional areas of the system and their layout and the locations of the various species and species mixtures, the system's boundaries, any human-made structures, and important natural features such as streams and ponds. Depending on the complexity of the system, it may be useful to include several other maps, each showing in greater detail some significant aspects of the system. For example, in a system with a complex annual cropping component, including a more detailed map of the layout of the area by crop type would be highly useful.

The report may conclude with a preliminary discussion of how system structure relates to system dynamics and functioning. This discussion can include ideas about the emergent qualities of the system, how the system is connected to the local economy and social structure, and which elements of the system contribute to its sustainability and which do not.

Variations and Further Study

1. Map the same agroecosystem each month during the growing season, or each year for several years, to learn how the system changes with time. Do the changes follow a pattern? Are they "natural" or human designed and intentional? If the latter is true, what is the logic of the farmer's strategy with respect to changing the structure and content of the agroecosystem over time?

2. Map a large agroecosystem or farming region with the assistance of aerial photos.

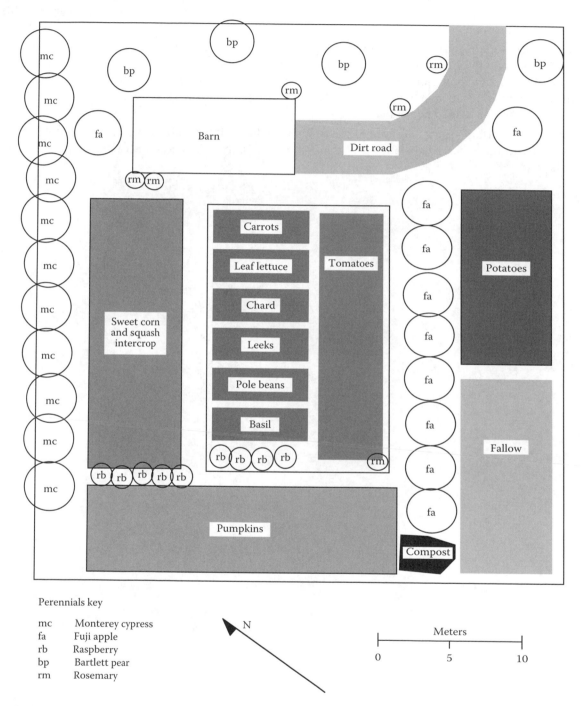

Figure 18.1
Example of an agroecosystem biodiversity map.

3. If an aerial photo and a geographical information system (GIS) are available, scan the aerial photo of the agro-ecosystem into the GIS system. Then actual coordinates (from a reference point) of locations on the photo can be established by ground surveys and entered into the system to create an electronic spatial database for the agroecosystem. The data can be used to visualize the system as a set of layers; each layer can contain a different type of information.

4. Use global positioning system (GPS) technology instead of constructing a transect grid. GPS devices, which give their precise location in the form of latitude and longitude, are now relatively inexpensive. If available, they can facilitate the mapping of larger systems. They can be used to determine absolute position in conjunction with a topographic map, and they can aid in the construction of a "virtual grid" because of their ability to show location relative to a reference point.

5. Map other important aspects of the agroecosystem, including soil type, pH, slope, water flux, yields, and nutrients. If a GIS base map has been created (see Variation #3), then these features can be mapped as electronic overlays.

Overyielding in an Intercrop System

Background

How densely can a crop be planted without adversely affecting the yield per unit of land? Agronomists have asked and experimentally answered this question for a variety of crops. As long as moisture or a critical nutrient is not limiting, a farmer can plant at a known ideal density and expect the maximum possible yield for a particular crop or variety. Beyond that density, plant–plant interference begins to become more significant, and overall yield declines.

It is possible, however, to overcome the planting-density barrier by planting a mixture of crops, called a polyculture or intercrop. In an intercrop, a variety of types of interference can prove beneficial to the crop plants, overcoming the effects of any interspecific or intraspecific competition that may come from close plant–plant spacing (Chapter 16 of *Agroecology: The Ecology of Sustainable Food Systems*). When two or more crop types are planted together, there is greater potential for mutualistic interference, complementary habitat modification, complementary nutrient use, and niche overlap (Chapters 14 and 17 of *Agroecology: The Ecology of Sustainable Food Systems*). Further, a crop mixture may show greater pest or weed resistance than either crop grown in monoculture.

Every particular mixture of crops will have its unique ecological dynamics. These dynamics can be studied, and their effects on yield determined. Specifically, the agroecologist is interested in whether a particular intercrop combination is capable of producing a higher yield than monocultures of the component crops grown on the same area of land. This capability is known as *overyielding* (Chapter 16 of *Agroecology: The Ecology of Sustainable Food Systems*). When overyielding occurs, intercropping may be an economically viable way of reducing dependence on external inputs and thus contributing to sustainability. Certain types of intercrops, such as the corn–bean–squash combination planted extensively in Mesoamerica, have been shown to overyield, often very significantly. Intercrop systems in general, however, need further study.

Textbook Correlation

Investigation 14: The Population Ecology of Agroecosystems (Niche Diversity and Overlap)
Investigation 16: Species Interactions in Crop Communities
Investigation 17: Agroecosystem Diversity

Synopsis

Two crops are planted separately (in monoculture) and together (in an intercrop), and then the yields of each treatment are measured to determine if the intercrop can produce a yield advantage relative to the monocrops. (Optionally, other data are collected to compare other ecological aspects of the two types of systems or to examine the mechanisms of interaction; see "Coordination with Other Investigations" and "Variations and Further Study" sections.)

Objectives

- Investigate the ecological dynamics of an intercrop.
- Compare intercropped systems and monoculture systems.
- Apply the land equivalent ratio (LER) as a way of determining the possible yield advantage of an intercrop.

Procedure Summary and Timeline

Week 1

- Sow seeds (or set out transplants) in experimental plots.

Weekly after week 1

- Measure various ecological factors (optional).

At crop maturity

- Measure harvest biomass (continuing weekly as necessary).

After last harvest

- Analyze data and write up results.

Timing Factors

This investigation requires careful and extensive planning (see "Advance Preparation" section for more information). Before deciding to conduct this investigation, it is important to determine if the local growing season combined with the timing of the period of academic instruction will allow adequate time for the development of the crop prior to collecting yield data. The shorter the growing season, the more restricted the options.

Coordination with Other Investigations

This investigation lends itself to extensive elaboration and modification, as described in Variations and Further Study. It can also be combined with several other investigations (see the list below), the field setup being used for related but distinct studies (an example is Investigation 15, which is written to make use of the intercropping field setup). This investigation, therefore, may be undertaken as a centerpiece of the laboratory section. If you want the investigation to assume this role, you should plan on having the crops mature well into the period of academic instruction, so that time is available for the measurement of ecological factors prior to harvest.

The following investigations, or variations of them, can be carried out in conjunction with this investigation:

Investigation 2: Light Transmission and the Vegetative Canopy

Investigation 3: Soil Temperature

Investigation 4: Soil Moisture Content

Investigation 11: Comparison of Arthropod Populations

Investigation 12: Census of Soil-Surface Fauna

Investigation 14: *Rhizobium* Nodulation in Legumes

Investigation 15: Effects of Agroecosystem Diversity on Herbivore Activity

Materials, Equipment, and Facilities

Uniform agricultural field (at least 15 m × 20 m; but a smaller-scale garden area may be used if necessary)

Tractor with implements appropriate to conditions of the field

Soil amendments (if necessary)

Irrigation system (if necessary)

String or tape and stakes for marking plot borders

Seeds or transplants of two crop plants

Scales suitable for weighing harvested crops to nearest 0.1 g

Soil temperature probes (optional)

Photometers (optional)

Soil core sampler (optional)

Drying oven (optional)

Advance Preparation

This investigation requires extensive advance preparation, most of which will probably occur before the beginning of the semester or quarter in which the lab is taught. The plots must be planted well ahead of time so that the material is ready for harvest during the time of the study.

- Select two crops for the investigation. Good candidates include sweet corn and squash, sweet corn and beans, lettuce and broccoli, and many others depending on the location.
- Make a field available for planting. Minimum size is approximately 15 m × 20 m; the ideal size (to maximize microclimate variation) is as large as is practical. The field should be relatively uniform (i.e., in slope and exposure) throughout. *Note*: If a farm-scale field is not available, the investigation can be carried out at a garden scale (see "Variations and Further Study" section).
- Map out the division of the field into blocks and plots. Each block will have two monocrop plots and an intercrop plot, but their arrangement within the block should be randomized (see the basic map in Table 19.1).
- Determine the dimensions of the field and its subunits. First, select a monocrop planting density appropriate to each chosen crop. This should be close to the density recommended for each crop. Second, based in part on the chosen planting densities, determine row spacing. Row spacing should be a minimum of 1 m from the middle of the crown to the middle of the other crown and should be the same in all plots. Third, assuming that plots will be

TABLE 19.1
Possible Randomized Plot Layout

Block I	Block II	Block III	Block IV
A monocrop	B monocrop	B monocrop	AB intercrop
B monocrop	AB intercrop	A monocrop	A monocrop
AB intercrop	A monocrop	AB intercrop	B monocrop

a minimum of five rows wide and that rows will run lengthwise through each block, determine how wide each plot (and block) must be (the minimum is 5 m), allowing also for an unplanted space between blocks if desired. Fourth, based on this figure, determine how wide the entire field must be (four times wider). Fifth, decide on the field's other dimension (the minimum is 5 m per plot × 3 plots deep = 15 m). If the intercrop setup is to be used for other investigations, the field should be somewhat larger than the minimum. This may mean, for example, using seven rows per plot and making the length of the blocks equal to or greater than the width of the field as a whole (28 m wide × 28 m deep).

- Prepare the field for planting. Depending on the climate, the chosen crops, and the condition and history of the field, preparation may include some combination of cover-cropping, cultivation, irrigation, and incorporation of soil amendments. Preparation may have to begin as early as 8 months prior to projected harvest. For example, in a dry-summer region, preparing the field for a late summer planting may involve the following: (1) plant cover crop in early spring, (2) mow and disc cover crop in late spring, (3) irrigate field in early summer to germinate weed seeds, (4) disc emerged weeds, (5) allow field to dry, (6) irrigate and disc once more, and (7) bed up rows.

- Based on the current weather conditions, predicted weather, time to crop maturity, and desired beginning of data taking (harvest), select a date for sowing the seed or transplanting seedlings. If any of the optional measurements of ecological factors described in "Variations and Further Study" section will be carried out, you will also want to factor in the desired timing of these measurements, some of which must occur prior to crop maturity and harvest. For some combinations of intercrops, one crop may need to be sown or transplanted before the other.

- Decide on a planting density for the intercrop plots. The overall density can range from one equal to that of the most dense monocrop (i.e., substitution density) up to one twice as dense as the monocrop (addition density). Base this decision on hypotheses about the complementarity of the two crops being used. (It is possible to use more than one intercrop plot per block, each with a different planting density; see "Variations and Further Study" section.)

- Sow the seeds (or set out transplants) in their appropriate plots (see Table 19.1). In the intercrop plots, sow or transplant so that different crop plants will alternate along a row as shown in Figure 19.1. In both monocrop plots and intercrop plots, allow for thinning of seedlings to achieve the desired density.

- Set up an irrigation system (if necessary for supplementing rainfall) and irrigate the sown seeds and/or transplants.

- Mark plot boundaries with suitable materials (e.g., poles and string). Label each plot.

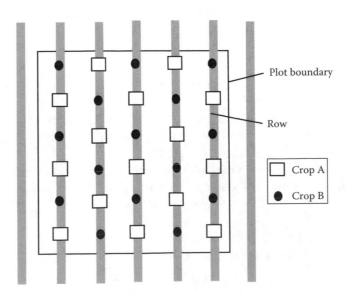

Figure 19.1
Example of plant arrangement in the intercrop plots.

Ongoing Maintenance

In addition to any necessary thinning of seedlings, the plots will require regular weeding/cultivation and irrigation until the crops are harvested for the determination of biomass (irrigation may be unnecessary, of course, if rainfall is sufficient for proper development of all crop species). It may also be necessary to control certain pests; for example, if gophers are a problem locally, they should be trapped. After the lab section is under way, weeding and pest control duties can be done by the investigation teams.

Investigation Teams

Form four teams of four to six students each. Each team will be responsible for collecting data from a set of three plots we will call a block. A block is made up of one monoculture of crop A, one monoculture of crop B, and one polyculture (intercrop) of the two crops. See the following logistics map:

Team 1	Team 2	Team 3	Team 4
Block I	Block II	Block III	Block IV
1 monocrop A	1 monocrop A	1 monocrop A	1 monocrop A
1 monocrop B	1 monocrop B	1 monocrop B	1 monocrop B
1 intercrop AB	1 intercrop AB	1 intercrop AB	1 intercrop AB

Procedure

Setup

Setup of the investigation occurs before the lab section is underway (see "Advance Preparation" section).

Data Collection

The process by which yield data are collected will depend on the types of crops planted. If both crops are types that can be harvested just once, such as broccoli and lettuce, all the harvesting and data collection can occur during one lab period. If, however, one or both crops are indeterminate producers, such as pole beans or squash, harvest and data collection may be spread over a period of many weeks. Further, if other ecological factors besides yield are being studied (see "Variations and Further Study" section), then data on these factors will be collected in the weeks prior to final harvest.

For a determinant crop, harvest should occur when a majority of the crop is of market size. For an indeterminate crop, the first harvest should occur when a sizable portion of the crop has reached market size and before any of the crop is overmature.

The following steps describe the collection of yield data for a one-time harvest in one block (three plots). For crops that require multiple weeks of harvest, these steps are repeated as necessary.

1. Delimit plot edges from plot centers. To eliminate edge effects, only data from plot centers will be used to calculate yields. In each plot, both outside rows are considered edges, along with a consistent (e.g., 0.5 m) portion at the ends of each central row. Make sure edge areas are the same size in each plot and well marked with string or other material.

Intercrop Datasheet					
Block: I Crop A: Lettuce Crop B: Broccoli					
	Harvested Biomass				
	Harvest Week 1; Date: 9/10, kg	**Harvest Week 2; Date:**	**Harvest Week 3; Date:**	**Harvest Week 4; Date:**	**Total, kg**
Crop A in monoculture	45	—	—	—	45
Crop B in monoculture	39	—	—	—	39
Crop A in polyculture	29	—	—	—	29
Crop B in polyculture	21	—	—	—	21

Figure 19.2
Example of a Completed Intercrop Datasheet.

2. Harvest the crop from the monoculture plots.

   a. If material from edge areas is to be harvested (i.e., for consumption by investigators), keep it separate from material from the central areas.

   b. Collect material in bags or other suitable containers that can be labeled and used to carry the material to the lab or weighing location. Label each container with group number, block, plot, and crop.

   c. Harvest the portion of the plant that would be normally marketed. Examples include sweet corn (full ear), squash (fruit and stem), lettuce (entire aboveground portion of plant), and broccoli (central stalk and florets). (But note that measuring total biomass of the crop, not just the harvested part, is an equally valid way of determining LER; see Variations and Further Study, #10.)

3. Harvest the crop from the intercrop plot. Harvest as described earlier, keeping each crop in separate containers.

4. Measure the marketable biomass (e.g., bean pods, corn ears, and lettuce heads) harvested from each plot. Record the data on the Intercrop Datasheet (Figure 19.2).

5. Repeat these steps in subsequent weeks if one or both of the crops are an indeterminate producer, adding up weekly yields for each crop and plot.

Data Analysis

Did the intercrop plots, on average, yield more crop on a per-land-unit basis than the monocultures of their constituent crops? To answer this question, we use the simple evaluative tool called the LER (Chapter 17 of *Agroecology: The Ecology of Sustainable Food Systems*). The LER is given by the formula

$$LER = \sum \frac{Yp_i}{Ym_i}$$

where
   Yp_i is the yield of each crop in polyculture (intercropped)
   Ym_i is the yield of each crop in monoculture

Intercrop LER Worksheet Block: I			
	Yield in Polyculture (Yp), kg	**Yield in Monoculture (Ym), kg**	**Partial LERs** $\dfrac{\mathbf{Yp}_i}{\mathbf{Ym}_i}$
Crop A: lettuce	29	45	0.64
Crop B: broccoli	21	39	0.54
			$\text{LER} = \sum \dfrac{Yp_i}{Ym_i} = 1.18$

Figure 19.3
Example of a Completed Intercrop LER Worksheet.

Each ratio of Yp/Ym for a particular crop is called a partial LER; added together, they constitute the total LER. A total LER greater than 1 indicates a yield advantage for the intercrop.

The following steps describe how to calculate a LER for the data from one block:

1. Transfer the yield data from the intercrop datasheet onto the Intercrop LER Worksheet.
2. For each row, divide the figure in the polyculture column by the figure in the monoculture column. The result, the partial LER for that crop, is entered in the rightmost column.
3. Add the two partial LERs to derive the overall LER and enter this figure in the last row of the chart.

Figure 19.3 shows an example of a LER calculated for the data in Figure 19.2. In this example, the calculated LER of 1.18 indicates slight overyielding for the intercrop. The figure of 1.18 can be interpreted to mean that 1.18 acres of the two monocultures would have to be planted to equal the yield on 1 acre of the intercrop.

When actual data are analyzed, each team determines a LER for its block, and these figures are averaged to determine an overall LER.

Write-Up

The following are some suggestions for writing up a report on the results of the investigation:

- Present a summary of the data in tabular or graphic form.
- Discuss the LER results and their significance.
- Discuss how the results may apply to actual farming situations or point to beneficial changes in design and/or management.
- Discuss variation between blocks.
- Evaluate the sustainability of the intercrop system.
- Present and discuss the results of any ecological factor measurements (described in "Variations and Further Study" section) and relate them to the LER results.

Variations and Further Study

Measurement of additional factors is highly recommended because of the investment of time and space required for this study and because additional data may be used to make inferences about the ecological dynamics responsible for overyielding (or lack thereof).

The following are suggestions for expanding the investigation by measuring additional factors; each one can be incorporated into the investigation singly or in combination with others.

1. Measure differences in the light environment of the three types of plots to assess the relative efficiency of the capture of sunlight. Measurements can include percent cover, relative rate of light transmission, or leaf area index (see Investigation 2). Light measurements can be taken at various stages in the growth of the plots, but if data are taken once, it should be just prior to the first measurement of harvest biomass.

2. Measure differences in the soil temperature of the three types of plots. The basic procedure is as follows: Measure soil temperature both at the surface and at a depth of 5 or 10 cm at midday at a number of different locations within each plot and repeat these measurements at regular intervals until the first measurement of harvest biomass. (See Investigation 3 for more information on methodology.) Certain crops might benefit from the lower soil temperatures likely to be found in the intercrop plots. Cooler temperatures might also mean slower ecological processes, such as those involving nutrient uptake and organic matter breakdown.

3. Measure differences in soil moisture among the plots. At several random locations within each plot, take a soil core sample down to a standard depth of 15 cm. Then weigh the soil from each sample, dry it in an oven at 105°C for 24 h, and weigh it again to determine the amount of water lost (as percent soil moisture). In some intercrops, soil moisture may remain at higher levels than in monocrops due to increased soil cover or be lower because of higher water use.

4. Measure differences in pest infestation or herbivore damage. Each type of crop will have a different kind of potential pest, and each is best counted in a particular manner at a particular time (e.g., corn earworm is best counted when the corn is harvested for biomass measurement; cabbage worm on broccoli is best counted just prior to harvest). The numbers of one pest species should be counted on the species' host both in the monocrop and in the intercrop. In some intercrops, pest infestation can be lower due to "masking" or other effects such as increased presence of predators on the pest. (See also Investigation 15, which can be carried out as a separate investigation using this experimental setup.)

5. Analyze the architecture of the intercrop system in order to describe resource partitioning. Compare the development, heights, leaf covers, and root structures of the crops making up the intercrop to determine (1) how well the plants might divide up the available resources (light, soil moisture, and soil nutrients) and (2) if each plant species might modify the system environment to the benefit or detriment of the other crop.

6. Measure differences in weediness among the plots. If you choose to measure this parameter, stop weeding all the plots at a certain point in crop development or never weed at all following sowing of the crops. Then at some later date (before measuring harvest biomass for the first time), set up quadrats and measure weed biomass and/or numbers of individual weed plants. In certain intercrops, weed growth may be diminished due to allelopathic effects or increased canopy cover.

It is also possible to modify the setup of the investigation so as to increase its complexity. Two possible modifications are as follows:

7. Within each block, set up intercrop plots with more than one planting density. Since yield in an intercrop is sensitive to density, having a range of densities will improve the chances of finding a density at which overyielding occurs. If possible, use three different densities: (1) a low density equivalent to the density of one of the monocrops (i.e., substitution density), (2) a high density equivalent to overlaying the density of one of the monocrops on the other (i.e., addition density), and (3) a medium density midway between these two extremes. When data from the plots are analyzed, calculate a LER relative to each density plot. Of course, having more than one intercrop plot per block will require a proportionally larger field.

8. Plant three crops. With this setup, there will be three monocrop plots within each block and one three-crop intercrop. Data are collected in a manner similar to that of the basic setup, and there are three partial LERs instead of two. An ideal three-crop mixture to test is the corn–bean–squash intercrop traditionally grown in Mesoamerica.

A third type of variation involves adding treatment variables to the investigation. One such variation is as follows, but others are possible.

9. Add another four blocks of plots identical to the first four, except leave the four additional blocks unweeded. This variation in treatments is likely to produce interesting differences in yield both within weeded and unweeded blocks and between the two types of blocks. Increased weediness can add more competitive interference, but it also increases diversity and the potential for beneficial interactions. (It is possible, e.g., for an intercrop to suppress weed growth through allelopathy or shading just enough to maximize the weeds' benefit.) To take full advantage of using this treatment variation, consider also measuring a number of additional ecological factors as described in points 1–4 earlier. Because of the increased number of plots and blocks, this variation works well with two lab sections (one assigned to the weedy blocks, the other to the weeded blocks) or one large section.

A fourth type of variation is to add another dimension to the data collection and analysis.

10. Calculate LERs based on total plant biomass produced, not just the biomass of the marketable crop. This analysis allows for the agroecological value of biomass that can be returned to the system, increasing the organic matter content of the soil and adding back extracted nutrients. The data for this calculation can be collected after the final measurement of harvest biomass. If plots are small, all remaining plants and root systems can be pulled up and their wet mass measured. The harvest biomass totals can then be added in. If the plots are larger, the total biomass per plot can be estimated by sampling every 10th plant (or remaining part thereof), determining its wet biomass, multiplying by 10, and adding in the harvest biomass totals. In either case, dry biomass can also be measured by first drying the plant material at 60°C for 48 h.

11. Calculate an economic LER based on local market prices of harvested crops. This analysis has a very practical application: it helps determine if growing an intercrop might be as profitable or more profitable than growing the same crops in monoculture. Overyielding by itself does not assure greater profitability for a variety of reasons. Most importantly, total LER is a sum of two or more partial LERs, and so it is possible that a total LER greater than 1 can be made up of a high partial LER for a relatively low-profit crop and low partial LER for an economically valuable crop. Calculating an economic LER is straightforward. When the crops are being harvested and weighed, simply determine their total market value based on local prevailing per-pound, per-kilogram, or per-unit prices (use retail prices if the crops could be sold consumer direct, as at a local farmer's market). Then use the resulting gross income figures in the LER formula. Note that a favorable economic LER does not by itself indicate profitability, since it does not take into account the costs involved in growing and harvesting monocultures versus polycultures.

Finally, the investigation can be carried out on a smaller scale if a larger field is not available.

12. Set up the investigation in garden beds instead of a farm field. Ecological differences between the monocrop and intercrop treatments may not be as noticeable at this scale, especially with regard to insect populations and herbivory, but measurable differences in LER can be obtained nonetheless. Use a randomized block setup similar to that described for the larger-scale setup, with each "plot" being a smaller hand-dug garden bed. Each bed should be at least four plants wide and 10 plants long to allow for a central sampling area two plants wide and eight plants long.

Intercrop Datasheet					
Block: Crop A: Crop B:					
	Harvested Biomass				Total
	Harvest Week 1; Date:	Harvest Week 2; Date:	Harvest Week 3; Date:	Harvest Week 4; Date:	
Crop A in monoculture					
Crop B in monoculture					
Crop A in polyculture					
Crop B in polyculture					

Intercrop LER Worksheet Block:			
	Yield in Polyculture (Yp), kg	**Yield in Monoculture (Ym), kg**	**Partial LERs** $\dfrac{Yp_i}{Ym_i}$
Crop A:			
Crop B:			
			$LER = \sum \dfrac{Yp_i}{Ym_i} =$

Grazing Intensity and Net Primary Productivity

Background

In a grazing system, heterotrophic animals feed upon pasture or range vegetation and the human managers of the system harvest milk, meat, or fiber from the animals. In ecological terms, the primary production of the vegetation is the basis for the secondary production of protein-rich animal products (Chapter 2 of *Agroecology: The Ecology of Sustainable Food Systems*).

Grazing is a form of herbivory, a removal interference in which an animal consumes plant tissue (Chapters 11 and 13 of *Agroecology: The Ecology of Sustainable Food Systems*). Even though the immediate effect on the consumed plants is negative, the overall effect on the ecosystem may not be. Removal of certain amounts and types of plant material can stimulate the production of new biomass or even allow certain desirable plant species in the community to germinate or become more predominant. Carefully managed grazing, therefore, can actually lead to greater productivity. The converse, of course, is also true: overgrazing can greatly reduce primary production, cause undesirable changes in the species composition of the vegetation, and lead to soil erosion.

Considerable research has been carried out on grazing systems to try to determine ideal grazing intensities and patterns. Depending on the pasture system and type of grazing animal involved, the ideal grazing pressure may be light and frequent, heavy and infrequent, or at some intermediate level.

Textbook Correlation

Investigation 2: Agroecology and the Agroecosystem Concept
Investigation 13: Heterotrophic Organisms in Agroecosystems
Investigation 18: Disturbance, Succession, and Agroecosystem Management
Investigation 19: Animals in Agroecosystems

Synopsis

In a pasture or range ecosystem, four different grazing patterns/intensities are simulated by establishing clip plots and subjecting their vegetation to different levels of forage removal. After 7–8 weeks, the regrowth of vegetation is measured to provide estimates of aboveground net primary productivity (NPP) during the span of the study. This data is then applied to understanding how different intensities and patterns of grazing may impact a grazing ecosystem.

Objectives

- Observe the impact of a removal interference on an agroecosystem.
- Simulate the effects of different grazing intensities.
- Measure NPP over time.
- Link the carrying capacity concept to the concept of sustainability.

Procedure Summary and Timeline

Prior to week 1

- Locate an appropriate grassland, pasture, or rangeland ecosystem and lay out plots.

Week 1

- Apply simulated grazing to the plots and measure biomass removed from each plot.

Week 8 or 9

- Remove and weigh biomass from each plot and calculate NPP for each.

After data collection

- Analyze data and write up results.

Timing Factors

This investigation is dependent on outdoor conditions and the availability of a pasture or rangeland system for study. It is best carried out during a time of the year when the pasture is growing rapidly.

Materials, Equipment, and Facilities

Four or more manual grass clippers

Meter tape for laying out blocks and plots

Stakes and twine to mark plots

Paper bags for drying biomass

Scale with 0.01 kg precision

Drying oven

A grassland, pasture, or rangeland system

Advance Preparation

- Locate a pasture or range agroecosystem or a grassland ecosystem. If such a system is not available, a large lawn can be used. Rangeland vegetation that includes woody perennials (such as Great Basin sage scrub) is not appropriate for this investigation, unless shrub density is low enough to allow placement of plots between shrubs.
- Within the chosen system, locate a study site approximately 6 m × 6 m. If the chosen site is currently being grazed, use fencing to exclude animals from the study site well before the investigation begins.
- Within the study site, lay out four blocks, each made up of four 1 m × 1 m plots. Leave enough distance between the blocks to allow easy access to the plots. Mark the boundaries of the plots with stakes and string or other suitable boundary markers.

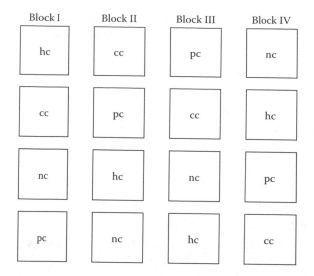

Figure 20.1
A possible randomized block layout of the plots.

- Draw a map of the blocks and plots and randomly distribute the four replications of the four different treatment types among the 16 plots, so that each block contains one replication each of the four treatment types. The treatments are named and described in the Procedure section. One possible randomized block layout is shown in Figure 20.1.

Ongoing Maintenance

If natural rainfall is not sufficient to maintain active growth of the ecosystem, some irrigation may be needed. If the system is normally grazed, animals must be kept away from the study plots with fencing or by other means.

Investigation Teams

Form four teams, each with three to five students. Each team will have responsibility for one block of plots, each of which is made up of one repetition each of the four treatments described in Setup and Initial Data Collection.

Procedure

Setup and Initial Data Collection

The following steps describe the procedure for one block of plots:

1. Label the plots in your block according to the layout map. Each plot will receive a different treatment, as described in the next four steps.
2. In the clear-clip (cc) plot, clip all aboveground biomass at ground level. Place the clippings in a marked paper bag.
3. In the high-clip (hc) plot, clip all aboveground biomass at a height of 10 cm above the ground level. Place the clippings in a marked paper bag.

4. In the patchwork-clip (pc) plot, clip half the aboveground biomass at ground level and leave the other half unclipped. This can be done by clipping every other 10 cm strip across the plot. Place the clippings in a marked paper bag.

5. In the no-clip (nc) plot, do nothing. This is the control plot.

6. Dry each paper bag full of clippings at 70°C for 48 h.

   Note: If the volume of fresh biomass from any of the plots overwhelms the available drying facilities, the dry mass of these clippings can be calculated by drying a smaller sample. The following steps describe how do to this:

   a. Find the fresh mass of all the clippings from the plot.

   b. Separate out approximately 10% of the clippings and find the fresh mass of this sample.

   c. Put the sample in a marked paper bag and dry it at 70°C for 48 h.

   d. Find the mass of the dried sample.

   e. Use the following equation to calculate the dry mass of the original pile of clippings from the plot:

   $$\text{Dry mass of all clippings} = \frac{\text{Fresh mass of all clippings} \times \text{dry mass of sample}}{\text{Fresh mass of sample}}$$

   This calculated dry mass is used instead of the measured dry mass in the first part of step #7.

7. Once dried, weigh the biomass from each plot to the nearest 0.01 kg. Record the data in the "Biomass removed initially" row of the Grazing Intensity Datasheet (Figure 20.2). Because the plots are all one meter square, the data can be expressed as kg/m^2.

8. Note the biomass removed initially from the **cc** treatment. Because it represents all the aboveground biomass from the plot, this value equals the original biomass (OB) for that plot. If we assume that all four plots began with virtually identical OBs, then this value also equals OB for the other three plots. Copy the value into each cell of the "Original biomass" row of the datasheet.

9. Calculate the "Starting biomass" for each plot—the aboveground biomass remaining in the plot after it is clipped:

   a. The starting biomass of the cc treatment is essentially equal to 0, which is already printed on the datasheet.

   b. The starting biomass of the hc treatment is equal to the OB of the plot minus the amount of biomass removed from the plot by clipping:

   $$\text{hc starting biomass} = \text{OB} - \text{Biomass removed initially}$$

   Perform this calculation and enter the result in the appropriate column of the "Starting biomass" row of the datasheet.

   c. The starting biomass of the pc treatment is also equal to the OB of the plot minus the amount of biomass removed from the plot by clipping:

   $$\text{pc starting biomass} = \text{OB} - \text{Biomass removed initially}$$

   Perform this calculation and enter the result in the appropriate column of the "Starting biomass" row of the datasheet.

   d. The starting biomass of the **nc** treatment is equal to the OB of the plot. Transfer the value for OB into the **nc** column of the "Starting biomass" row of the datasheet.

Grazing Intensity Datasheet

Date of Initial Clipping: April 13, 2015

Date of Final Clipping: June 08, 2015

Elapsed Time: 56 Days

Block: B

Team: B

	Treatment			
	cc	hc	pc	nc
Original biomass (OB) (kg/m²)	3.09	3.09	3.09	3.09
Biomass removed initially (kg/m²)	3.09 =OB	1.12	1.56	0
Starting biomass (kg/m²)	0	1.97	1.53	3.09 =OB
Biomass removed at the end = Final standing biomass (kg/m²)	0.78	2.53	2.51	3.53
NPP (kg/m²) (=Biomass removed at end – Starting biomass)	0.78	0.56	0.98	0.44

cc starting biomass = 0.
hc starting biomass = OB – Biomass removed initially.
pc starting biomass = OB – Biomass removed initially.
nc starting biomass = OB.

Aggregation of Data from All Blocks

	Block				Mean NPP of Treatment
	A	B	C	D	
cc NPP (kg/m²)	0.73	0.78	0.74	0.82	0.77
hc NPP (kg/m²)	0.51	0.56	0.60	0.64	0.58
pc NPP (kg/m²)	0.89	0.98	1.03	0.92	0.96
nc NPP (kg/m²)	0.42	0.44	0.49	0.50	0.46

Figure 20.2
Example of a Completed Grazing Intensity Datasheet.

Final Data Collection

The following steps are carried out 7 or 8 weeks after the initial removal of biomass from the plots:

1. Clip all the biomass in each plot at ground level. Be careful not to include soil or roots in the clipped matter.
2. Place each plot's clippings into a separate labeled paper bag.
3. Dry the bags of clippings at 70°C for 48 h. If the volume of fresh biomass from any plot is too large to dry as a whole, use the sampling procedure described earlier in Setup and Initial Data Collection, step #6.
4. After drying, find the mass of each pile of clippings to the nearest 0.01 kg. (Or, if the sampling procedure was used, calculate the dry mass of each pile of clippings.)
5. Record the data in the appropriate columns of the "Biomass removed at end" row of the datasheet.

Data Analysis

1. Calculate NPP for each plot in your team's block. NPP is equal to the biomass produced over a particular time period. In other words, NPP = ending biomass – beginning biomass. (For the purposes of this investigation, the portion of NPP used to increase root biomass is being ignored.)

   a. For each plot, subtract the value for "Starting biomass" from the value for "Biomass removed at end."

   b. Record each result in the appropriate column of the "NPP" row of the datasheet.

2. Share your team's NPP data with the other teams.

3. Collect other teams' NPP data. Record them in the table in the lower part of the datasheet. Find the mean NPP for each treatment and enter each figure in the right-hand column of the lower table.

Write-Up

The following are some suggestions for reporting on the results of the investigation:

- Present the data in a form that allows easy comparison of the results from each treatment.
- Discuss the observed relationships between intensity of grazing and productivity.
- Explore possible relationships between carrying capacity and sustainability of grazing and pasture systems.
- Discuss the possible agroecological significance of the plant–herbivore interaction.

Variations and Further Study

1. Investigate the effects of simulated herbivory on crop plants. Using a crop such as broccoli, remove varying amounts of biomass from plants when the heads are beginning to form and measure the effect on the biomass of the harvested heads.

2. Add treatment plots in which the simulated grazing is repeated at regular intervals (e.g., every 2 weeks) throughout the study period. Collect, dry, and weigh the biomass removed at each clipping, add up the amounts from each clipping to arrive at a figure for total biomass removed, and use this figure to determine NPP. Compare the results with the data from the plots that are clipped only once.

3. Investigate the effects of the different simulated grazing regimes on species composition and diversity, both of which are important determinants of forage quality. Each time biomass is removed from the plots, separate it by species and find the biomass of each species. Calculate NPP for each species and compare the results from different treatments. It is also possible to calculate a species diversity index for each treatment using biomass instead of numbers of individuals.

Grazing Intensity Datasheet				
Date of Initial Clipping: Block:				
Date of Final Clipping: Team:				
Elapsed Time:				
	Treatment			
	cc	**hc**	**pc**	**nc**
Original biomass (OB) (kg/m²)				
Biomass removed initially (kg/m²)	=OB			
Starting biomass (kg/m²)				=OB
Biomass removed at the end = Final standing biomass (kg/m²)				
NPP (kg/m²) (=Biomass removed at end – Starting biomass)				

cc starting biomass = 0.
hc starting biomass = OB – Biomass removed initially.
pc starting biomass = OB – Biomass removed initially.
nc starting biomass = OB.

Aggregation of Data from All Blocks					
	Block				**Mean NPP of Treatment**
	A	**B**	**C**	**D**	
cc NPP (kg/m²)					
hc NPP (kg/m²)					
pc NPP (kg/m²)					
nc NPP (kg/m²)					

21

Effects of Trees in an Agroecosystem

Background

Trees form part of the agricultural landscape in many parts of the world. When trees are intentionally retained or planted on land used for crop production or grazing, the resulting systems are considered examples of agroforestry (Chapter 18 of *Agroecology: The Ecology of Sustainable Food Systems*). Trees, or woody perennials more generally, can have a great many beneficial ecological impacts on the overall productivity and stability of an agroecosystem. The objective of most agroforestry systems is to optimize these beneficial effects and minimize the negative. But thorough agroecological knowledge of trees' ecological roles and influences are necessary to accomplish this goal.

Trees are capable of altering dramatically the conditions of the ecosystem of which they are part (Chapter 18 of *Agroecology: The Ecology of Sustainable Food Systems*). Belowground, a tree's roots penetrate deeper than those of annual crops, affecting soil structure, nutrient cycling, and soil moisture relations. Aboveground, a tree alters the light environment by shading, which in turn affects humidity and evapotranspiration. Its branches and leaves provide habitats for an array of animal life and modify the local effects of wind. Shed leaves provide soil cover and modify the soil environment; as they decay, they become an important source of organic matter and nutrients.

Because of these influences, trees in agroecosystems can be a good foundation for developing the emergent qualities of a more complex agroecosystem. Solar energy capture can be enhanced; nutrient uptake, retention, and cycling can be improved; the entire system can be maintained in dynamic equilibrium. The permanency that trees provide in terms of resources and microsites can stabilize populations of both pests and their predators. Overall, such interactions can help reduce the dependence of the system on outside inputs.

Textbook Correlation

Investigation 18: Disturbance, Succession, and Agroecosystem Management

Synopsis

In this open-ended investigation, various ecological factors are measured along transects that begin at the trunk of a tree and range beyond the canopy. The data collected are used to characterize the localized influences of the tree, providing useful information on the potential role that trees integrated into agriculture might play in establishing more sustainable design and management.

Objectives

- Investigate the ecological impacts of trees.
- Gain experience measuring ecological factors along a transect.

Procedure Summary and Timeline

Prior to week 1

- Identify an appropriate tree for study.

Week 1

- Lay out transects and begin collecting data.

Weeks 2 and 3

- Continue data collection as necessary.

After data collection

- Analyze data and write up results.

Timing Factors

This investigation can be carried out almost any time of the year that the soil is not frozen, waterlogged, or covered with snow; however, the ideal time is during a period of active plant growth, when trees' impacts are most evident.

Materials, Equipment, and Facilities

Stakes and string for defining transects

Meter tape

Marker

Single tree away from other trees

(Other materials will depend on which factors are measured)

Advance Preparation

- Locate an appropriate tree for study. It should be a lone tree that exerts an obvious influence on its immediate surroundings, such as a tree in a meadow, a tree in a park or garden, or a tree left in a farming system of some kind. The tree should be mature, so that its effects on the area will have accumulated over time. Ideally, the vegetation under the tree will be rather uniform so that the tree's effects will not be confounded by the influences of a diverse vegetation.
- From among the options listed in Data Collection, choose an appropriate set of ecological measurements to make along each transect. Consider the time and equipment that is available, students' prior experience, and pedagogical goals.
- Collect the materials required for each type of data collection.

Ongoing Maintenance

None required.

Investigation Teams

Form four teams, each with three to five members. Each team will be responsible for one of the transects established at the study tree.

Procedure

Setup

1. Lay out four transects from the tree trunk outward, one in each compass direction. The lengths of the transects should be equal to the approximate diameter of the tree's canopy. Because the transects begin at the center of the canopy, this length means that the transects will extend well beyond the canopy in each direction. Define the transects clearly with stakes and string.

2. Determine an appropriate distance between data collection points along the transects. This will depend on the lengths of the transects. The target number of data collection points is 24, so begin by dividing the lengths of the transects in meters by 24. For example, 12 m transects can have 24 points in 0.5 m increments, 10 m transects can have 20 points in 0.5 m increments, and 14 m transects can have 23 points in 0.6 m increments.

3. Using a meter tape, mark each data collection point along all four transects with tape or a permanent marker.

General Notes on Data Collection

Each suggested type of data collection along the transects is described in the sections following. In some cases, you will need to refer to the methods described in other investigations.

If you are measuring all four of the soil factors that require collecting a sample, it is possible to collect just two soil samples from each location rather than four. You will need separate samples for determining bulk density and soil moisture content, but either sample can then be used to measure (in this order) nutrient levels and percent organic matter content.

The datasheet for this investigation is intended for processed or averaged data. For some types of measurements, this means you will record the "raw" data in another location (the lab notebook or a copy of the appropriate datasheet) and then transfer the meaningful, processed data to this datasheet. If data are collected on different days, be sure to record the date on which each set of observations was made.

Regardless of which ecological factors are measured along the transects, it is a good idea to map the tree's canopy. In its simplest form, the map is a rough picture of the shade the canopy would cast if the sun were directly overhead, showing how the shade overlaps each transect (see Figure 21.1). To create the map, mark the location on each transect where the canopy edge is directly overhead and measure the distance from this location to the tree's trunk. Additional maps can be created that show how the shade cast by the canopy moves depending on season and time of day.

Measuring Light Transmission

Using a dual-sensor, photosynthetically active radiation light meter, measure the light level at each data collection point on the transect, following the basic methodology outlined in Investigation 2, Data Collection step #2. Note the following considerations:

1. Rather than place the transect sensor at ground level, put it on top of a pole that is just slightly longer than the height of the vegetation along the transect. This modification is necessary because the object is to measure the shading effect of the tree, not the shading of the surrounding vegetation as well.

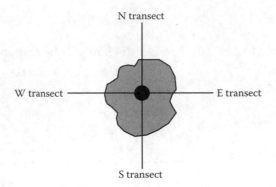

Figure 21.1
Example of a simple canopy map.

TABLE 21.1
Example of a Light Transmission Datasheet

Transect Location	Light Level along Transect			Full-Sunlight Level	
	Meter Reading	Corrected Light Level	% of Mean Full-Sunlight Level	Meter Reading	Variance from Mean
1	489	484	5.3%	9104	+5
2	513	467	5.1%	9145	+46
24	8963	9096	100%	8966	−133
			Mean → 9099		

2. The object of this investigation, in contrast to that of Investigation 2, is to measure changes in the shading effect of the tree along the transect, not determine an average light transmission value along the transect. The data, therefore, are handled differently, and a datasheet different from that in Investigation 2 is required. Table 21.1 shows an example of the type of datasheet needed. Use this model to create one in your lab notebook.

3. Follow the procedure immediately following to collect and process the light transmission data. (If a dual-sensor meter is not available, use the alternate method described in Investigation 2.)

    a. At each point along the transect, take a light-meter reading, enter it in the datasheet, and then immediately switch to the other sensor to take a full-sunlight meter reading.

    b. When the light levels have been measured at every point along the transect, calculate the mean full-sunlight level.

    c. Find the difference between each full-sunlight meter reading and the mean. Record the result as a positive number if the reading is higher than the mean; record it as a negative number if it is lower than the mean.

    d. For each transect light-meter reading, subtract the amount of the variance if the variance is positive, and add it if it is negative. The resulting corrected light level reflects variations in the full-sunlight reading caused by passing clouds and other atmospheric changes.

    e. Divide the corrected transect light level by the full-sunlight level mean and then multiply by 100 to obtain the percentage of the full-sunlight mean.

4. Transfer the data for percentage of the full-sunlight mean to the appropriate locations in the light transmission column of the Tree Effects Datasheet (Figure 21.2).

Tree Effects Datasheet

Transect: West-Pointing

Date: August 17, 2016

Tree Species: *Cupressus macrocarpa* (Monterey Cypress)

Time: 13:00 h

Data Collection Point	% Light Transmission	Soil Water Content (θ_m)	Soil Surface Temperature (°C)	Organic Matter Content (%)	Soil Bulk Density (g/cm³)	Litter Layer Depth (mm)	Understory Height (cm)	Understory Diversity (# sp)
1	2.1	0.30	17.9	4.5	1.42	21	5.1	1
2	1.9	0.29	18.1	4.6	1.45	21	7.2	2
3	2.2	0.27	18.4	4.5	1.49	22	6.3	1
4	2.3	0.25	18.3	4.3	1.44	24	5.9	3
5	5.2	0.24	18.1	4.2	1.49	22	7.3	2
6	3.2	0.26	18.5	3.8	1.45	20	4.1	2
7	4.6	0.29	18.7	4.0	1.42	19	7.9	2
8	7.9	0.27	19.2	3.9	1.43	15	4.9	3
9	5.8	0.32	18.9	3.7	1.53	13	11.7	5
10	10.6	0.28	19.3	3.4	1.45	8	14.4	4
11	15.3	0.25	20.2	3.3	1.47	6	20.4	7
12	42.3	0.21	21.1	3.2	1.53	2	29.5	10
13	100	0.20	24.0	2.8	1.65	1	32.3	9
14	100	0.20	25.1	2.7	1.71	0	31.5	11
15	100	0.18	27.4	2.7	1.73	0	29.8	12
16	100	0.17	29.8	2.8	1.69	1	30.9	10
17	100	0.19	33.0	2.5	1.74	0	30.3	12
18	100	0.17	34.9	2.6	1.75	0	32.5	13
19	100	0.16	35.2	2.7	1.77	1	31.6	12
20	100	0.16	35.5	2.6	1.79	0	32.0	11
21								
22								
23								
24								
25								

Figure 21.2
Example of a Completed Tree Effects Datasheet.

Measuring Soil Moisture

Using the method described in Investigation 4, measure the soil water content at each data collection point along the transect. Note the following considerations:

1. For the purposes of this investigation, measurement of moisture content within one stratum—0–15 cm below the surface—is adequate (but measure the moisture in other strata if desired).

2. Use the Soil Moisture Datasheet (the same one used in Investigation 4) for recording the raw data. If measuring only at the 0–15 cm depth, use the rows labeled 0–15 cm and ignore those labeled 15–30 cm. Also ignore the column for can number, since sample location and can number will be identical. Do not bother to calculate a mean soil water content.

3. After calculating the soil water content (θ_m) for each data collection point, transfer this value to the soil water content column of the Tree Effects Datasheet.

Measuring Soil Temperature

Using the method described in Investigation 3, measure the soil temperature at each data collection point along the transect. Note the following considerations:

1. It is adequate to measure soil temperatures at the surface only, but other depths can be measured if desired.
2. Record the soil temperature readings directly on the Tree Effects Datasheet.

Measuring Organic Matter Content of the Soil

Using the method described in Investigation 5, measure the organic matter content at each data collection point along the transect. After calculating the percent organic matter content for each data collection point, record these data on the Tree Effects Datasheet.

Measuring Nutrient Content of the Soil

Using the method described in Investigation 5, measure the levels of nitrogen, phosphorus, and potassium present in the soil at each data collection point along the transect. Record the data in your lab notebook or in a supplemental datasheet of your own making.

Measuring Bulk Density of the Soil

Using the method described in Investigation 5, measure the bulk density of the soil at each data collection point along the transect. After calculating the bulk density of the soil at each data collection point, record these data on the Tree Effects Datasheet.

Measuring Depth of the Litter Layer

At each data collection point along the transect, use a ruler to measure the depth of the litter layer to the nearest millimeter. Record the measurements on the Tree Effects Datasheet. See "Variations and Further Study" section for more sophisticated ways of measuring changes in the litter layer along the transects.

Measuring Height of Understory Vegetation

Follow these steps at each data collection point along the transect:

1. Lay a meterstick perpendicular to the transect line, with the 50 cm point set on the line.
2. At the 0, 25, 50, 75, and 100 cm points of the meterstick, use another meterstick to measure the height of the nearest plant (if the understory vegetation is a crop, measure crop plants only).
3. Find the mean of the five plant heights and record this figure in the appropriate column of the Tree Effects Datasheet.

Measuring Understory Vegetation Species Diversity

This measurement should be carried out only if the vegetation under the tree is not a crop.

Follow these steps at each data collection point along the transect:

1. Center a 50 cm × 50 cm quadrat over the data collection point.
2. Count the number of species present inside the quadrat and record this figure in the appropriate column of the Tree Effects Datasheet.
3. If desired, count the number of individuals of each species and calculate the Shannon diversity index of the quadrat (see Investigation 10 for a description of how to calculate the index).

Other Possible Measurements

- Rate of moisture infiltration in the soil
- Soil tilth
- Soil color
- Soil aggregate size
- Abundance, species diversity, and biomass of earthworms
- Biomass of understory vegetation
- Diversity index of understory vegetation
- Root volume in soil cores taken at 0–30 cm and 30–60 cm (see "Variations and Further Study" section)

Data Analysis

1. Determine how each ecological factor changes along your transect.
2. Look for relationships between different factors and between each factor and the tree's canopy.
3. Share the data for your transect with other teams and gather together the data for the other transects.
4. Compare the data from all four transects to look for effects of slope, prevailing wind, and exposure.

Write-Up

The following are some suggestions for reporting on the results of the investigation.

- Present the data for each ecological factor in a graphical form showing how the factor changes along each transect.
- For each factor, determine where the tree exerts an effect and where the effect begins to diminish.
- Interpret the various patterns shown in the data.
- Discuss the effects of slope, prevailing winds, and exposure, as indicated by differences among the four transects.
- Discuss which of the tree's effects are potentially positive and which are potentially negative from the standpoint of agroforestry.
- Explain how the positive effects of trees can be maximized in agroecosystems.

Variations and Further Study

1. Do this study with a deciduous tree species and collect data in both the summer and in the early spring before the tree leafs out. Use the data to determine which influences are direct and which are residual.

2. Carry out the investigation on several same-species trees of varying age and compare their influences on the ecosystem.

3. Plant a crop along each of the four transects, either as seed or transplants, and compare responses over time.

4. Set up small quadrats along the transects for measuring the dry mass per unit area of the litter layer.

5. Set up litterfall catchers along the transects, as described in Investigation 6, to measure the input of litter from the tree over time.

6. Investigate how some of the ecological factors may change over time: soil temperature over a 24 h period; soil moisture over a week, month, or season; and light transmission over a 12 h daylight period.

7. Get some idea of the structure of the belowground components of the system by taking soil cores with a sharp-edged coring device at two or more depths (such as 0–30 cm and 30–60 cm) at each data collection location on a transect and by measuring the volume of the roots from each core as described in Investigation 8.

8. Measure any other ecological factor along the transects that may be affected by the tree.

Soil Moisture Datasheet

Sampling Date:

Agroecosystem:

Sample Location	Can #	Sample Depth (cm)	Mass of Sample Can (g) A	Mass of Can and Wet Soil (g) B	Mass of Can and Dry Soil (g) C	Mass of Water Lost (g) $B - C$	Mass of Dry Soil (g) $C - A$	Soil Water Content θ_m (g Water/1 g Dry Soil) $\dfrac{B - C}{C - A}$
1	1	0–15						
	2	15–30						
2	3	0–15						
	4	15–30						
3	5	0–15						
	6	15–30						
4	7	0–15						
	8	15–30						
5	9	0–15						
	10	15–30						
6	11	0–15						
	12	15–30						
7	13	0–15						
	14	15–30						
8	15	0–15						
	16	15–30						
							Total	
						Mean soil water content, 0–15 cm		
						Mean soil water content, 15–30 cm		

Tree Effects Datasheet

Transect: Date:

Tree Species: Time:

Data Collection Point	% Light Transmission	Soil Water Content (θ_m)	Soil Surface Temperature (°C)	Organic Matter Content (%)	Soil Bulk Density (g/cm³)	Litter Layer Depth (mm)	Understory Height (cm)	Understory Diversity (# sp)
1								
2								
3								
4								
5								
6								
7								
8								
9								
10								
11								
12								
13								
14								
15								
16								
17								
18								
19								
20								
21								
22								
23								
24								
25								

Section V

Food System Studies

Investigation 22

On-Farm Energy Use

Background

Agriculture, in essence, is the human manipulation of the capture and flow of energy in ecosystems. Humans use agroecosystems to convert solar energy into particular forms of biomass—forms that can be used as food, feed, fiber, and fuel. All agroecosystems—from the simple, localized plantings and harvests of the earliest agriculture to the intensively altered agroecosystems of today—require an input of energy from their human stewards in order to capture and convert into useful form the energy provided by the sun. This input is necessary in part because of the heavy removal of energy from agroecosystems in the form of harvested material. But it is also necessary because an agroecosystem alters many natural ecosystem processes. Humans must intervene in a variety of ways—manage noncrop plants and herbivores, irrigate, cultivate soil, and so on—and doing so requires work.

The agricultural "modernization" of the last several decades has been largely a process of putting ever greater amounts of energy into agriculture in order to increase harvestable yields. But most of this additional energy input comes directly or indirectly from nonrenewable fossil fuels, a form of industrial–cultural energy (Chapter 20 of *Agroecology: The Ecology of Sustainable Food Systems*). This reliance on industrial–cultural energy over biological–cultural energy is a major factor in shifting our energy "balance sheet" into the red: for many crops, we invest more energy than we get back as energy in food. Because of this unfavorable return on our energy investment, coupled with our dependence on nonrenewable energy sources, agriculture cannot be sustained into the long-term future without fundamental changes in how we view and use energy (Chapter 20 of *Agroecology: The Ecology of Sustainable Food Systems*).

A good first step toward using energy in agriculture more sustainably is to simply become more aware of how energy is used to grow food and to consider how local and renewable biological–cultural energy sources might begin to replace nonrenewable industrial–cultural energy sources.

Textbook Correlation

Investigation 20: Energetics of Agroecosystems

Synopsis

Farmers are interviewed to gather information on how they use energy on their farms and to learn more about how they perceive energy use. Questions are asked concerning energy types, costs, renewability, and future availability.

Objectives

- Learn interviewing skills.
- Explore the issues of energy availability, cost, and sustainability.
- Gain an understanding of farmer perceptions of on-farm energy use issues.

Procedure Summary and Timeline

Prior to week 1

- Contact farmers and arrange interview times.

Week 1

- Carry out interviews.

Week 2

- Do follow-up interviews based on questions raised in the first interview.

After interview

- Analyze and discuss findings and write up report.

Timing Factors

This investigation can be carried out during any time of the year but might be best accomplished during a time when farmers are less occupied with farm activities, such as during the winter, between planting and harvest, or while the farmer is selling at a farmers' market.

Materials, Equipment, and Facilities

Notepad and paper

Clipboard

Audio recorder (optional)

Advance Preparation

- Contact local farmers and ascertain their interest in being interviewed. When you talk to a farmer, make it clear that you value his or her knowledge and point of view and want to learn from them. Try to represent a variety of farm sizes, types of practices, types of crops, and geographic areas in your choice of initial contacts. Attempt to obtain commitments from as many farmers as there are teams.
- Arrange times for interviews. Make it clear that an interview may take an hour or more. Obtain permission from each interviewee to record the interview.

Ongoing Maintenance

None required.

Investigation Teams

Form teams with two to four members each. Each team will be responsible for interviewing one farmer (or farming team).

Procedure

Data Collection

1. Before the interview, learn as much as possible about the farmer being interviewed: crops grown, farm size, types of practices used, problems related to the geographic area, and so on. This knowledge will help you form more intelligent questions.

2. Review the datasheet and make a list of questions to ask after gathering the basic information in the datasheet. Keep in mind the following:

   a. Your questions should be adapted to the individual you are interviewing (the questions you would ask a small-scale organic farmer, e.g., are somewhat different from those you would ask a larger-scale conventional farmer).

   b. Many farmers are used to being treated by researchers and extension agents as receivers of information, not sources.

   c. You will get more interesting information and establish better rapport if you let the farmer's point of view control the agenda.

3. Discuss how you, as a team, will conduct the interview. Will you take turns asking questions? Will you each have responsibility for different areas of questioning? (see Investigation 23 for interviewing tips)

4. Meet the farmer for the interview with a notepad and pen or pencil, the On-Farm Energy Use Datasheet (Figure 22.1), a clipboard, and (if possible) an audio or voice recorder.

On-Farm Energy Use Datasheet

Farm: Sunburst Farm

Date: December 2, 2015

	Used?	Used for?	Rank Importance	Rank Cost	Comments, etc.
Gasoline and diesel	✓	Running tractors, trucks	1	2	
Propane and natural gas	✓	Flame weeder	7	8	
Electricity	✓	Well pump	2	3	
Synthetic fertilizers	✓	Fertilizing supplementally	6	5	
Synthetic pesticides and herbicides	✓	When other controls do not work	8	6	
Human labor	✓	Cultivation, harvest, transportation, etc	4	1	Hires workers seasonally
Animal labor					Would consider using
Compost and manure	✓	Primary soil amendment	3	7	From on-farm and local sources
Seeds and transplants	✓		5	4	
Wind power					Would like to use
Hydropower					

Figure 22.1

Example of a Completed On-Farm Energy Use Datasheet.

5. After introducing yourselves and your purpose, begin by gathering basic, concrete information about the farmer's use of energy. Use the datasheet as a guide.

   a. Go through each type of energy on the datasheet, asking the farmer if he or she uses that type of energy and what it is used for. Make notes of any additional information the farmer offers through this process. In particular, note the farmer's reaction when you mention pesticides, fertilizers, compost, seeds, etc., as energy sources. Explain why these inputs are considered energy inputs and again note the farmer's view.

   b. Tell the farmer you would like him or her to rank each of the energy sources they use by importance, meaning how much of the total annual energy use that type of energy makes up. Read the list of sources again and ask the farmer to assign each one a rank, with 1 being the most important energy source.

   c. Repeat this process for a ranking by cost.

6. Ask additional questions about energy use, guided by what you have discovered by going through the datasheet. Here are some possible questions:

   a. Which types of energy would you like to use less of? Why?

   b. Which types would you like to use more of? Why?

   c. Do you foresee any future availability or cost problems for any of the energy sources you use?

   d. What factors are preventing you from using energy in the way you would like?

   e. Are your organic-matter inputs—compost, manure, etc.—free or purchased?

   f. Have you made any changes in the way you use energy on the farm in recent years?

   g. Do you see any potential for using wind, solar, or hydropower on your farm? How exactly would you use these energy sources and what fossil-fuel energy uses would they replace?

Initial Interview Review

Meet after the interview as a group to discuss your findings. Discuss the following:

- Does the farmer understand the difference between solar or natural system energy and human-derived or cultural energy? The difference between renewable and nonrenewable energy?

- How dependent is the farmer on nonrenewable energy?

- What would need to be done to help move the farmer in the direction of using more renewable and sustainable forms of energy?

- What are the farmer's subjective views of what limits changes in how he or she uses energy? How do you size up the limiting factors objectively?

- What additional information, if any, should you ask for in a follow-up interview?

Final Interview

Do another interview if you think it would be productive and the farmer is agreeable.

Data Analysis

1. If the interview was recorded and time allows, transcribe the interview.

2. Summarize the most important information in the transcript (or interview notes).

3. Meet with other teams to compare your experiences and data.

Write-Up

Use the following questions as guidelines for writing up a report:

- What is the pattern of energy use on your farm? How does this pattern compare to those of the farms examined by other teams?
- How do the uses of renewable and nonrenewable energy compare?
- What factors seem to act as barriers to farmers changing their current patterns of energy use?
- How vulnerable economically is the farmer to spikes in the cost of fossil-fuel-based energy sources?
- Does the farmer view energy use patterns in terms of sustainability?
- What is your definition of sustainable energy use in agriculture?

Variations and Further Study

1. By consulting relevant literature, obtain figures for the energy content in actual kilocalories of the different forms of energy inputs to agriculture. Return to the farmers who are agreeable and obtain data on more exact amounts of each form of energy used on the farm system. The units will vary depending on the energy type, from gallons to hours to kilograms. Convert these amounts to a common energy content unit, such as kilocalories. Obtain a total figure for energy inputs on the farm, and using the same conversion data, calculate the energy production of the farm. Determine the ratio of cultural energy inputs to energy outputs for the farm.

2. Consult sources that discuss how to determine the "intrinsic" energy content in industrial–cultural inputs to the farm. This includes the energy costs of refining, transporting, and merchandising gasoline or diesel fuels; the energy required to manufacture, transport, and maintain a tractor; and the energy needed to produce seeds. Investigate how to incorporate such energy expenditures into the energy use calculations describe earlier.

3. Interview local advocates for the transition to renewable and sustainable sources of energy, with a focus on learning what sources are available or may soon become available, what is needed for farmers to more readily adopt them, and what the primary limitations to their adoption might be.

On-Farm Energy Use Datasheet					
Farm:					Date:
	Used?	Used for?	Rank Importance	Rank Cost	Comments, etc.
Gasoline and diesel					
Propane and natural gas					
Electricity					
Synthetic fertilizers					
Synthetic pesticides and herbicides					
Human labor					
Animal labor					
Compost and manure					
Seeds and transplants					
Wind power					
Hydropower					

Investigation

Farmer Interview

Background

Farmers—the hands-on managers of agroecosystems—have enormous stores of knowledge about what works and what does not and why. They understand the locality in which they farm, the variations and extremes in its weather, the pests that must be contended with, the crops that respond best, the soil and what it needs to remain productive. Regardless of how sustainable an individual farmer's practices are, his or her knowledge is an important resource, and his or her concerns and point of view are something the agroecologist must take into account. For these reasons, a central tenet of agroecology is that local, farmer-based knowledge is a key starting point in the movement toward sustainability (Chapter 25 of *Agroecology: The Ecology of Sustainable Food Systems*).

A farmer's knowledge and practices, however, must also be understood within the larger context of the food system. Farming is an economic activity, and so a farm's place in the web of food production, distribution, and consumption relationships affects everything a farmer does on the farm. It matters who buys a farm's products, what price the farmer gets, and how far the food travels to get to the consumer's table.

Many farmers feel helpless in the face of globalization and the increasing power of the food processing, transporting, marketing, and retailing "middlemen." Some farmers, however, are shifting their marketing strategies in ways that let them retain control over their economic fates. Many farmers moving in this direction are making their farms the primary building blocks of alternative food systems that eschew globalization and work to re-create more direct connections between the growers and consumers of food (Chapter 25 of *Agroecology: The Ecology of Sustainable Food Systems*).

Textbook Correlation

Investigation 22: Converting to Ecologically Based Management
Investigation 23: Indicators of Sustainability
Investigation 25: Culture and Community in the Remaking of the Food System

Synopsis

A farmer (or farming team) is interviewed to learn about the farming practices, knowledge, motivations, major challenges, role in the food system, and goals for the future. The information gained from the interview may be used for later collaborative problem solving.

Objectives

- Learn interviewing techniques.
- Understand the human element of agriculture.
- Investigate farming as "ethnoscience."
- Gain a baseline of local knowledge.
- Explore the role that local farmers play as components of regional and global food systems.
- Take steps toward transforming the relationship between farmers and agroecological researchers into one that is more participatory, with information flowing in both directions.

Procedure Summary and Timeline

Prior to week 1

- Contact interview candidates and arrange interview times.

Week 1 (2, 3)

- Conduct interviews.

After interviews

- Transcribe interviews and write up reports.

Timing Factors

This investigation can be completed in a relatively short amount of time, with little preparation. It can be done any time of the year, but it may be best to avoid times that would be especially busy for local farmers.

Coordination with Other Investigations

This investigation may be combined with Investigation 18. If the two are combined, however, the mapping exercise should be carried out first, before the investigators receive much input from the farmer.

Materials, Equipment, and Facilities

Several willing farmers

Audio recorders (optional)

Advance Preparation

- Contact local farmers and ascertain their interest in being interviewed. When you talk to a farmer, make it clear you value his or her knowledge and point of view and want to learn from them. Try to represent a variety of farm sizes, types of practices, types of crops, and geographic areas in your choice of initial contacts. Attempt to obtain commitments from as many farmers as there are teams.
- Arrange times for interviews. Make it clear that an interview may take an hour or more. Obtain permission from each interviewee to record the interview.

Ongoing Maintenance

No maintenance is required.

Investigation Teams

Form two-person interview teams, each of which will interview a different farmer. If the number of willing farmers is limited, team size can be increased to three; teams larger than three persons may have greater difficulty establishing rapport with the interviewee.

Procedure

Data Collection

1. Before the interview, learn as much as possible about the farmer being interviewed: crops grown, farm size, market conditions, problems related to the geographic area, and so on. This knowledge will help you form more intelligent questions.
2. Make a list of questions to ask. Keep in mind the following:
   a. The questions should be adapted to the individual you are interviewing (the questions you would ask a small-scale organic farmer, e.g., are somewhat different from those you would ask a larger-scale conventional farmer).
   b. Many farmers are used to being treated by researchers and extension agents as receivers of information, not sources.
   c. The interview has two basic objectives: (1) learning about the farmer's practices, problems, motivations, marketing strategies, and farming history at a descriptive level and (2) understanding the logic and knowledge that underlies the farmer's practices and goals. These dual objectives mean that for every practice listed in "Areas to Investigate," you should learn *what* the farmer does and *how* and *why* he or she does it.
   d. You will get more interesting information and establish better rapport if you let the farmer's point of view control the agenda. It may be best to find out about certain practices or rationales with indirect questions, rather than direct ones. A direct question such as "Do you use integrated pest management?" can be interpreted as "Do you control pests the right way or the wrong way?" and is best avoided.
   e. As the interviewee gets to know and trust you during the interview, you may be able to ask more probing or potentially sensitive questions.

   Areas to Investigate
   - Pest management
   - Maintenance of soil health and organic matter management
   - Cultivation
   - Weed management and use of "good weeds"
   - Cover cropping and use of fallow cycle
   - Crop combining and polyculture
   - Farm diversity and integration with natural vegetation
   - Use of animals
   - Erosion control
   - Use of energy

- Soil moisture management and irrigation
- Use of trees and other perennials
- Connection with the local community
- Marketing and economics of the farming enterprise

Possible Interview Questions

- How long have you been farming? What led you into farming?
- Why is farming important to you? What is your greatest source of satisfaction?
- What do you grow? Why?
- What is the most difficult aspect of farming for you?
- How do you manage pests? What is your biggest pest problem?
- How often do you cultivate?
- How do you fertilize your soil?
- Where do you get your seeds?
- How has your soil changed during the period you have farmed here? How do you keep your soil healthy?
- Do you plant cover crops?
- What do you see as your biggest challenge in the coming year?
- What are you doing to keep your land productive over the long term?
- What role do you see the natural or less disturbed areas of your farm playing?
- Have you tried any new techniques in the past few years? What are they? Have they worked or not? Why?
- Is your farm significantly different from others in the area? In what ways?
- Have you shifted, or considered changing, your marketing strategy in recent years?
- What do you see as the underlying cause of recent changes in the economics of agriculture?
- What do you see yourself doing on your farm in 10 years?
- To whom do you sell your crops? What proportion of your farm's produce remains in the local region?
- Have you heard of community-supported agriculture (CSA)? Have you considered creating or joining a CSA?

3. Discuss how you, as a team, will conduct the interview. Will you take turns asking questions? Will you each have responsibility for different areas of questioning?

4. Conduct the interview, using an audio recorder if possible (if the interview is not recorded, take careful notes). The questions you have listed are only a starting point; they are only tools for prompting the interviewee to disclose the information you seek. Be flexible; do not mechanically march through the list of questions.

5. If the interview was recorded and time allows, transcribe the interview.

6. Summarize the most important information in the transcript (or interview notes).

Write-Up and Presentation of Data

A suggested approach is to divide the report into two parts. In the first part, present what you have learned from the interviewed farmer, using a descriptive mode. In the second part, shift to an analytical mode. Discuss the farmer's practices, strategies, and problems from an agroecological perspective, using the following questions as guidelines:

- Why is it important to take into account the local-farming-knowledge base of an area?
- How sustainable are the farmer's practices?

- What role could agroecological researchers play in helping the farmer shift to more sustainable practices?
- What might agroecological researchers learn from the farmer's knowledge or practices? Could some of the practices be adopted more widely in the area? What elements of the farmer's knowledge should be verified or documented by agroecological research?
- How might the problems and challenges identified by the farmer be solved?
- What would have to change (in the local market, in government policy, etc.) for farmers in the area to be motivated to shift to more sustainable practices?
- What opportunities are available for the farmer to become part of an alternative food network?

Either before or after reports are written, it may be instructive to hold an informal roundtable discussion in which interview teams share their experiences and findings. Consider the possibility of delivering copies of the completed reports—after instructor critique and revision—to the farmers.

Variations and Further Study

1. Combine each team's report into a larger and more comprehensive regional report. Such an undertaking could be the basis of a senior thesis project.

2. Return for a second round of interviews with each farmer. Investigators may have built some trust during the first interview, allowing more in-depth questions and follow-ups during a second interview.

3. Use the interviews as a basis for constructing an "oral history" of each farmer, farming family, or farming community. Focus the interview questions on such matters as intergenerational transfer of knowledge and technique, change over time on the farm and in the community, and how new methods are learned and shared.

4. Focus the interview around identifying the farmer's major challenges and problems. Learn how the farmer perceives the causes of and potential solutions to these problems.

5. Focus the interview around the economics of the farming enterprise and the farm's role in the food system.

24

Local Food Market Analysis

Background

In most developed countries, consumers have available a wide variety of farm products and fresh produce. But to what extent does diversity in the marketplace reflect diversity in farmers' fields? Since most conventional farms grow only a few crops, usually in extensive monocultures, it would seem that market diversity and ecological diversity do not necessarily go hand in hand.

Agricultural sustainability will depend on there being economic incentives for ecological diversity in the field rather than disincentives. A first step in this process is simply understanding the linkages between the market and farmer's practices. Agriculture is both an economic activity and an ecological one. A farm that is not economically viable cannot last for long. Nevertheless, if economic forces alone determine what is produced and how it gets produced, sustainable practices tend to be abandoned in favor of those that will yield short-term profit and an immediate return on investment.

Textbook Correlation

Investigation 25: Culture and Community in the Remaking of the Food System
Investigation 26: From Sustainable Agroecosystems to a Sustainable Food System

Synopsis

A number of local retail produce sites (farmers' markets, groceries, and supermarkets) are surveyed to determine what is available in the local market, where it comes from, how it is grown, and what it costs. This information is then used to construct a picture of the local food market and shed light on how market forces may affect agricultural practices.

Objectives

- Explore the effect of market forces on farming operations.
- Understand aspects of the local food system.
- Link the market system to sustainability.

Procedure Summary and Timeline

Before week 1

- Identify potential survey sites.

Week 1 (2, 3)

- Survey individual sites where produce is sold.

After surveys are complete

- Collect together and analyze data from each survey team and put together a picture of the local food market.

Timing Factors

This investigation can be carried out whenever fresh local produce is being harvested and delivered to market.

Materials, Equipment, and Facilities

Clipboards

A representative mixture of produce retail sites, including farmers' markets, independent groceries, and supermarket chain stores

Advance Preparation

- Make a preliminary list of possible retail sites for students to survey.
- Decide how to define "local" for your area. The definition can be bioregional (i.e., by valley or drainage basin), economic (i.e., how the food market is geographically segmented), or geographic (i.e., by county or urban region). At a farmers' market, all produce for sale should be considered local, even if it was grown in a different region.
- Discuss the purpose of the investigation and the issues it will help address, using textbook Chapters 25 and 26 of *Agroecology: The Ecology of Sustainable Food Systems* as guides.
- Form teams (as described in Investigation Teams) and assign a retail site to each team or allow teams to choose their sites. Keep a number of sites "in reserve" for teams who may encounter uncooperative produce managers at their initial sites.
- Make an appropriate number of copies of the datasheet.

Ongoing Maintenance

None required.

Investigation Teams

Form two-person survey teams, each of which will survey the produce available at one retail site. If the number of potential sites is limited, team size can be increased to three.

Procedure

Data Collection

1. If your assigned or chosen retail site is a grocery store or supermarket, find out if the origin or grower of each produce item is given on signs (only a few stores do this). If the information is not given, contact the produce manager of the store. Ask if it will be possible to talk with him or her to find out where each produce item in the store comes from. In some cases, the produce manager will simply not know the origin of many items, because they are purchased from distributors who provide little or no information on the sources of their produce. If this is the case, do not worry, because the inability to track down sources is itself significant information. If your site is a farmers' market, try to locate the farmer or other people who actually participated in the growing of the food. People hired only to sell the food will not know very much.

2. Survey the fresh produce for sale at the site.

   a. List each produce item and its variety on a copy of the Food Market Datasheet (Figure 24.1). Use as many copies of the datasheet as needed. You may want to group items by type or by grower.

   b. For each produce item, answer the following questions:

      i. What is the source of the item? Where was it grown? Be as specific as possible, listing farm name (if known) and state or country (if known).

      ii. Was the item grown locally or was it imported? Use the definition of "local" decided on previously.

Food Market Datasheet

Community: Lawrenceburg, MI

Survey Date: June 17, 2015

Market Name: Sam's Market

Type of Market Location: ☑ Independent Grocery   ☐ Supermarket Chain

☐ Farmers' Market

☐ Community-Supported Agriculture

Produce Type	Variety	Source	Local or Imported?	Organic or Conventional?	Retail Price
Lettuce	Oak leaf	Green Valley Farm	Local	Organic	$1.29/head
Lettuce	Romaine	Green Valley Farm	Local	Organic	$1.29/head
Lettuce	Romaine	MegaCrop, Inc.	Imported	Conventional	$1.19/head
Lettuce		MegaCrop, Inc.	Imported	Conventional	$.89/head
Tomato	Roma	Mexico	Imported	Conventional	$.79/lb
Tomato	Hothouse cluster	The Netherlands	Imported	Conventional	$3.49/lb
Tomato	Medium salad	Mexico	Imported	Conventional	$.99/lb
Bell pepper	Green	Unknown	Imported	Probably conventional	$1.49/lb

Figure 24.1
Example of a Partially Completed Food Market Datasheet.

    iii. Was the item grown organically or conventionally?

    iv. Was the item grown in a monoculture or in a mixture? How many other types of produce are grown at its farm of origin?

    v. What is the item's retail price?

A few grocery stores present all the necessary information on their produce labels. At most groceries and supermarkets, however, information on where and how the produce was grown must be obtained by talking with the produce manager. To optimize the time spent with the produce manager, list every produce item on the datasheet first. At a farmers' market, the information is easily obtained by talking with each farmer.

3. As possible, collect additional information about the following:

    a. Market intermediaries (i.e., "middlemen")

    b. Storage of produce (where, how, how long)

    c. Transport of produce

    d. Shortages, surpluses, unfavorable conditions, and other situations affecting supply and price

    e. Perceived consumer preferences

Data Analysis

The data can be analyzed in two separate phases. First, each team studies and summarizes the data it has collected on individual retail sites.

1. Look for patterns and trends in the data.

2. Summarize the data and put it into a form that can be easily presented to other teams.

3. Identify the issues that arose during the survey process. List questions to be raised.

    Then, teams gather together to present what they have found, integrate the data that has been collected, and begin to form a picture of the local food market. This process can range from brief and cursory to involved and lengthy.

4. Put together the data from all the teams in some reasonable manner.

5. Discuss the structure of the local food market, using the following questions as a framework:

    a. Approximately what percentage of the produce consumed locally is grown locally?

    b. What percentage is grown organically?

    c. How does the local market seem to affect local growers and the decisions they make about what to grow and how to grow it?

    d. Which aspects of the local food market contribute to sustainability of the food production system and which do not?

Write-Up

There are two basic options for reporting on the results of this investigation:

1. Individual or team reports can be prepared prior to the second, whole-class phase of data analysis.

2. Individual or team reports can be prepared after the second phase of analysis and incorporate the class findings.

In either case, the questions and suggestions in the previous section can serve as guidelines for what to include in the reports.

Variations and Further Study

1. Combine this investigation with Investigation 23. A practical way to do this is to focus on how the market affects individual farmers' decisions about what to grow and how to grow it.

2. If a farmers' market has just been established in the local area, focus this investigation on how it changes the local produce market and affects the farmers who participate in it.

3. Cooperate with a local farmer in studying the process of increasing diversity on the farm coupled with introducing a new crop into the local market.

4. Investigate how changes in consumer behavior during some recent time span have affected farmer practice and the local food system. Or look into how consumers, acting to become better food citizens, can cause positive changes in the local or regional food system.

Food Market Datasheet

Community: Survey Date:

Market Name:

Type of Market Location: ☐ Independent Grocery ☐ Supermarket Chain

☐ Farmers' Market

☐ Community-Supported Agriculture

Produce Type	Variety	Source	Local or Imported?	Organic or Conventional?	Retail Price

Section VI

Instructor's Appendix

Appendix A: Planning the Field and Laboratory Course

Laboratory courses in agroecology and related fields are taught in a great variety of situations and contexts around the world. The institutional context ranges from large universities with extensive facilities to modest field schools with no formal laboratories. They are located all over the world, in every type of climate where agriculture is practiced, from northern and southern temperate climates to arid regions and the humid tropics. Some courses last a weekend or two, others 10 or 15 weeks, and some a whole year. Some correspond with the local growing season, some overlap it at the end or the beginning, and others occur when the fields outside are covered with snow or water. The students taking the courses have a wide range of backgrounds; some have little formal training in science, and some have no practical experience in agriculture or gardening. Class size ranges from a handful of students to several hundred divided among sections.

This manual was written with the recognition that it might be used in any of these situations. The investigations include many that need no specialized equipment or facilities and some that can be carried out when the weather outside is not conducive to plant growth. Many of the investigations can be carried out in a variety of climates; some can be abbreviated if time is short; some can be simplified or done at a smaller scale if the suggested facilities or equipment is not available.

Despite this flexibility, the investigations had to be written on the basis of certain assumptions about the typical situation and context. These assumptions direct how the investigations are described and organized, but they do not prevent you from adapting the investigations' basic ideas and procedures to your needs.

One of the assumptions is that the investigations are carried out by a single class or lab section made up of about 12–25 students. The number of replicates in experimental procedures and the extensiveness of the analyses and comparisons are based on this minimum number of students. If your course includes more than one lab section or has more than about 25 students, you can of course modify the investigations to make use of the additional data-collecting capabilities; some suggestions for how to do this are given under "the Modifying Investigations section in this Appendix and the Variations and Further Study section of each investigation."

A second assumption is that the lab course or section meets once a week. With two exceptions, each investigation's timeline of procedures is based on this meeting-time interval. (The exceptions are two entomological investigations in which data must be collected 1 or 2 days after the initial setup.) Laboratory courses with other than a once-a-week schedule can nevertheless adapt many of the investigations to their unique situations.

A third assumption is that at least 2 h is available at each meeting time. Some of the setup and data-collection procedures will take 2 h or more to complete.

Equipment and Facilities

What do you need to carry out the investigations? Each investigation has its own set of necessary tools, equipment, and facilities; these are listed near the beginning of each investigation. In some cases, you can substitute a piece of equipment, a tool, or a facility with another. Instructors should review an investigation's equipment and facilities list before deciding to include it in their curriculum.

The basic needs for carrying out many of the investigations are listed in the following. If you lack anything listed, you may be able to work around it or find a substitute.

- Laboratory space—or an area that can serve as a laboratory. Most of the investigations need an area that can be used for weighing samples, sorting materials, and performing similar activities. It is recommended that the laboratory space also have a chalkboard or whiteboard and a sink and water source.
- Basic lab equipment, including scales, thermometers, meter sticks, and a drying oven. Some investigations do not require very much in the way of equipment; others require more specialized equipment than is listed here.
- Garden or field space for planting. Not all the investigations require planting space, but those that do assume that at least several good-sized garden beds can be employed. Some investigations are based on field-scale plantings, but can be adapted to the use of a smaller space.
- Basic gardening tools. For several investigations, you will need such tools as spades or trowels. The few investigations based on field-scale plantings require a tractor and basic implements. In some contexts, pest-control equipment (e.g., gopher traps) or irrigation equipment will be necessary.
- Materials for amending the soil. Before setting up the plantings required in some of the investigations, the soil may need to have amendments added. These may include composts or manures.
- Existing agroecosystems nearby. Many of the investigations involve studying agroecosystems that have already been set up for the purpose of growing food. These can be independent from the educational or research institution or a part of the institution's programs.

Many instructors will be fortunate enough to have all the items listed available at one site or several adjacent sites, in the form of a working farm or garden connected with the college or university, or an agricultural experiment station. If your situation is not so fortunate, you may have several other options. It may be possible to find planting space, lab space, or existing agroecosystems at independent farms in the area, at another research or educational institution, or even at a local elementary or secondary school with a large teaching garden.

As a planning aid, Table A.1 lists the types of equipment and facilities needed for each investigation. A check in the Specialized Equipment column does not necessarily indicate that expensive or unusual equipment is required.

Choosing Investigations to Match Your Goals

There are a number of possible goals in teaching a laboratory course in agroecology. These include, but are not limited to

- Providing opportunities for hands-on exploration and application of agroecological concepts
- Teaching agroecological and/or ecological field and lab techniques
- Providing practice and experience in manipulating data and doing statistical analysis
- Teaching how to carry out controlled scientific experiments
- Giving students direct experience in farm and agroecosystem management

TABLE A.1
Equipment and Facilities

	Basic Field and Lab Equipment and Lab Space	Specialized Equipment	Greenhouse or Lath House	Field or Garden Space for Planting	One or More Existing Agroecosystems
Effect of microclimate on seed germination	✓	✓			
Light transmission and the vegetative canopy	✓	✓			✓
Soil temperature	✓	✓			✓
Soil moisture content	✓				✓
Soil properties analysis	✓	✓			✓
Canopy litterfall analysis	✓				✓
Mulch system comparison	✓	✓			✓
Root system response to soil type	✓	✓	✓		
Intraspecific interactions in a crop population	✓			✓	
Management history and the weed seed bank	✓		✓		✓
Comparison of arthropod populations	✓				✓
Census of soil-surface fauna	✓				✓
Bioassay for allelopathic potential	✓	✓			✓
Rhizobium nodulation in legumes	✓	✓	✓		
Effects of agroecosystem diversity on herbivore activity	✓	✓		✓	
Herbivore feeding preferences	✓	✓			
Effects of a weedy border on insect populations	✓		✓	✓	
Mapping agroecosystem biodiversity	✓				✓
Overyielding in an intercrop system	✓	✓		✓	
Grazing intensity and net primary productivity	✓				✓
Effects of trees in an agroecosystem	✓	✓			✓
On-farm energy use					✓
Farmer interview					✓
Local food market analysis					✓

TABLE A.2
Characteristics of the Investigations

	Experimental	Descriptive/ Comparative	Statistical	Technique-Focused
Effect of microclimate on seed germination	✓		✓	
Light transmission and the vegetative canopy		✓		✓
Soil temperature		✓	✓	✓
Soil moisture content		✓		✓
Soil properties analysis		✓		✓
Canopy litterfall analysis		✓		
Mulch system comparison	✓		✓	
Root system response to soil type	✓		✓	✓
Intraspecific interactions in a crop population	✓			
Management history and the weed seed bank		✓		✓
Comparison of arthropod populations		✓		✓
Census of soil-surface fauna		✓		✓
Bioassay for allelopathic potential	✓		✓	✓
Rhizobium nodulation in legumes	✓		✓	
Effects of agroecosystem diversity on herbivore activity		✓		
Grazing intensity and net primary productivity	✓			
Herbivore feeding preferences	✓			
Effects of a weedy border on insect populations	✓			
Mapping agroecosystem biodiversity		✓		✓
Overyielding in an intercrop system	✓			✓
Effects of trees in an agroecosystem		✓		✓
On-farm energy use		✓		
Farmer interview		✓		✓
Local food market analysis		✓		

The investigations in this manual are all designed for satisfying the first goal in the list, but they vary in how much attention they give to the other goals. By examining Table A.2, you can get a general picture of what each investigation offers. Keep in mind, however, that with appropriate modification, any of the investigations can satisfy most of the goals listed.

The categories in Table A.2 relate to pedagogical goals as follows:

Experimental: Investigations described as experimental involve manipulation of conditions and control of variables. They may be performed in the lab or in the field. They expose students to more-or-less rigorous experimental method.

Descriptive/comparative: Investigations in this category involve measurement, description, or analysis of one or more existing agroecosystems or their parts. The goal of these investigations is to characterize or compare agroecosystems. They provide students with opportunities for studying how agroecosystems vary in space and time with regard to conditions, structure, function, diversity, and so on.

Statistical: Investigations are included in this category when they involve any statistical analysis more complicated than calculating a mean. These investigations offer students experience in performing statistical calculations and applying statistics to analysis of experimental results and quantitative measurements.

Technique-focused: Investigations designated as technique focused involve using an important agroecological laboratory, field, or analytical technique, such as measuring light transmittance, estimating root system volume, or calculating a land equivalent ratio or diversity index. These investigations acquaint students with techniques that can be applied broadly in agroecological study.

Choosing Investigations for Your Climate and Season

In many climates, periods of academic instruction do not correspond well with the local growing season, making agroecological studies difficult. Although the majority of the investigations in this manual are designed to be carried out during the growing season, there are a good handful that are independent of climate and season (see Table A.3). Moreover, a number of the investigations that do require conditions conducive to plant growth can be performed during the periods of transition between the growing season and the cold or dry season, such as during autumn or spring in temperate climates, early or late winter in Mediterranean climates, or around harvest time in seasonal wet–dry climates. Finally, some of

TABLE A.3
Dependence of Investigations on Outside Conditions

	Independent of Climate or Season	Carried Out during Growing Season
Effect of microclimate on seed germination	✓	
Light transmission and the vegetative canopy		✓
Soil temperature		✓
Soil moisture content		✓
Soil properties analysis	✓[a]	
Canopy litterfall analysis		✓[b]
Mulch system comparison		✓
Root system response to soil type	✓[c]	
Intraspecific interactions in a crop population		✓
Management history and the weed seed bank	✓[c]	
Comparison of arthropod populations		✓
Census of soil-surface fauna		✓
Bioassay for allelopathic potential		✓
Rhizobium nodulation in legumes	✓[c]	
Effects of agroecosystem diversity on herbivore activity		✓
Grazing intensity and net primary productivity		✓
Herbivore feeding preferences		✓[d]
Effects of a weedy border on insect populations		✓
Mapping agroecosystem biodiversity	✓	✓
Overyielding in an intercrop system		✓
Effects of trees in an agroecosystem	✓	✓
On-farm energy use	✓	
Farmer interview	✓	
Local food market analysis		✓

[a] Soil cannot be frozen.
[b] Leaves must be present on perennials.
[c] Soil cannot be frozen; greenhouse needed if cold outside.
[d] Need source of insects and plant material.

the investigations designed around an outdoor planting can be adapted to run in a greenhouse. Therefore, instructors teaching an agroecology laboratory course at any time of the year in any part of the world will have at least some investigations to choose from.

Among the investigations based on an outside planting or an existing agroecosystem, nearly all can be performed in any type of climate. You simply choose crop types suited to the local conditions or study the types of agroecosystems in the local area.

Scheduling the Investigations

Making up a schedule for the laboratory course can be a complex process. You must estimate the time required to complete setups and procedures, build in some flexibility to deal with the vagaries of the weather and unforeseen problems, use time efficiently, and take into account the timing of due dates for reports. To assist you in this process, each investigation includes a procedural timeline and a section on "Timing Factors."

In general, an investigation takes more than 1 week to complete, and the amount of work required each week for an investigation varies from 10 min of data taking or watering to 3 h of setup. Therefore, several investigations can be running concurrently, with students working on different phases of perhaps two, three, or four investigations during any one lab period. A typical 3 h lab period might be spent as follows:

Discussion of results of investigation A	0.25 h
Discussion of setup of investigation D	0.5 h
Setup of investigation D	1.5 h
Maintenance work on investigation C	0.25 h
Data collection for investigation B	1.5 h

If the lab course is being taught during a time of seasonal change—such as during the fall semester in a temperate climate—one of the prime considerations in scheduling is adapting the schedule to the progression of the seasonal changes and their effect on growing conditions. Much activity may have to be compressed into the part of the instructional period with the most amenable weather, with data analysis or indoor studies filling up the remainder.

If you intend to focus the lab course on investigations that take many weeks to complete, you may find that all the data analysis and lab report writing are concentrated in the final weeks of the course. This situation may leave students feeling overwhelmed, and the lack of earlier experience in these tasks may leave them feeling unprepared as well. To avoid this problem, it is recommended that you include a shorter investigation early in the course, so that students can gain experience working with data and writing a lab report, and get useful feedback about their performance.

Combining and Coordinating Investigations

Many of the investigations are interconnected and can be combined or integrated in a number of ways.

- The physical setup of one investigation can be used as the study site for another. The setup of monocrop and intercrop plots used for Investigation 19, for example, can provide the study site for several other investigations.
- A narrowly focused investigation can become an integral part of a broader, more multifaceted investigation. The former can retain its identity as a separate investigation or it can become a subpart of the latter. For example, Investigation 3 can become an integral part of Investigation 19.

TABLE A.4
Possible Relationships between Investigations

A	B	Role of B in Terms of A	Role of A in Terms of B
Light transmission and the vegetative canopy	Overyielding in an intercrop system	B offers A a set of existing agroecosystems to study	A can be considered an integral part of B
Soil temperature	Overyielding in an intercrop system	B offers A a set of existing agroecosystems to study	A can be considered an integral part of B
Soil temperature	Mulch system comparison	B offers A a set of existing agroecosystems to study	An abbreviated version of A is part of B
Soil moisture content	Overyielding in an intercrop system	B offers A a set of existing agroecosystems to study	A can be considered an integral part of B
Soil moisture content	Mulch system comparison	B offers A a set of existing agroecosystems to study	An abbreviated version of A is part of B
Comparison of arthropod populations	Overyielding in an intercrop system	B offers A a set of existing agroecosystems to study	A can be considered an integral part of B
Census of soil-surface fauna	Overyielding in an intercrop system	B offers A a set of existing agroecosystems to study	A can be considered an integral part of B
Rhizobium nodulation in legumes	Overyielding in an intercrop system	B can provide the setup for a field version of A	A can be considered an integral part of B
Effects of agroecosystem diversity on herbivore activity	Overyielding in an intercrop system	B provides A with the set of existing systems for study	A can be considered an integral part of B
Mapping agroecosystem biodiversity	Farmer interview	B may follow A on the same farm	
Effects of trees in an agroecosystem	[Various investigations]	Techniques from several investigations are part of A	
Local food market analysis	Farmer interview	If A follows B, B can be a source of useful data for A	

- An investigation may include exploration of a phenomenon or relationship that is the central focus of another separate investigation. For example, Investigation 7 involves measuring soil temperatures, which is also the subject of Investigation 3. In the case of this example, temperature measurement can be done in a relatively simple manner described in Investigation 7, or it can be done more thoroughly as described in Investigation 3, with the temperature data being used in both investigations.

- Investigations can be carried out in sequence, each maintaining its separate identity but the two together providing a much broader picture of the study subject. For example, a farm can be mapped in Investigation 18 and then the farmer can be interviewed in Investigation 23.

Many investigations with a potential for some kind of coordination with others (or a written-in connection) contain a section, "Coordination with Other Investigations," that describes the connections and suggests ways of coordinating the investigations. Table A.4 summarizes these relationships.

Modifying Investigations

To allow instructors to carry out the investigations with a minimum of worry about procedural details, the "how-to" parts of the investigations are fairly specific and detailed. This does not mean that there is only one way to set up an investigation and collect and analyze its data. The investigations are intended to be completely open to modification. Each is really an example of how to carry out one type of investigation, one way of exploring a particular concept, relationship, or interaction. At the end of each investigation, a section on "Variations and Further Study" gives suggestions for modifying the investigation.

One particular type of modification deserves note: Since the investigations are written for courses with a single group of 12–25 students, instructors teaching courses with more than that number of students, either in one class or in multiple sections, will need to modify the investigations appropriately. For larger single classes, this means simply adding more teams and study plots or replications. For multiple sections, the modifications may be less straightforward, especially for the experimental investigations. They might involve, for example, adding variables or treatments.

Accumulating Data from Year to Year

By their nature, the investigations in this manual are limited to studying changes over time in the short term. However, if the same agroecosystems or subsystems are studied year after year, it is possible to accumulate some meaningful longitudinal data. Anticipating this possibility will help you set up studies and collect data in ways that will provide the most useful data for future analysis. Investigations suited to studying long-term changes in the same systems include Investigations 5, 8, 10, 18, 22, and 23.

Running the Field and Laboratory Course

Making a lab and field course a good learning experience for students requires guidance, encouragement, coordination, and direction from an instructor or teaching assistant (TA). It is up to the instructor or TA to clearly communicate the goals and procedures of an investigation; to have gathered the necessary materials ahead of time; to have set up any necessary plantings; and to help students form groups, work together, and manage time effectively. This section is intended to assist instructors and TAs in carrying out these responsibilities.

Preparing and Maintaining the Investigation Setups

All of the investigations require at least some advance preparation (defined as tasks performed before students begin work on the investigation). In a few cases, advance preparation involves performing some task (collecting weed seeds, e.g.) up to a year before the investigation is actually carried out. In other cases, one or more crops must be planted several weeks to several months before data collection begins. To help instructors and TAs understand the preparation required for each investigation, a section on advance preparation has been included with each investigation.

Some of the preparation tasks may require quite a bit of work—cultivating, bedding up, and planting a relatively large field, for example. If you plan to carry out investigations that involve such extensive preparation, you will obviously need assistance. Ideally, the course will be taught in conjunction with some type of ongoing agricultural operation with staff to whom you can hand over such duties.

Depending on circumstances, instructors may choose to organize an investigation such that students complete some or all of the advance preparation tasks as part of their coursework. In other cases, it may be necessary to treat some of the tasks described in the investigation's "Procedure" section as advance preparation.

A number of the investigations require ongoing maintenance—such as irrigating and weeding fields or removing samples from drying ovens. In some cases, these tasks can be performed by students as an integral part of carrying out the investigation. In other cases, maintenance tasks must be performed when students are not likely to be available (e.g., 24 h after a lab meeting). For these latter investigations, it is important to make prior arrangements for having the tasks carried out by staff or student volunteers. As an aid for coordinating maintenance tasks, each investigation includes a section entitled "Ongoing Maintenance," which briefly describes what is needed for that investigation.

Working with Teams

The investigations are all designed to be carried out by students working in teams. A team is a group of two to perhaps eight students who work together closely in carrying out some discrete part of an investigation and collecting a set of data. In experimental investigations, a team will typically carry out either one replication or will collect data on one distinct type of setup. In other types of investigations, a team may have responsibility for collecting data in one of several different agroecosystems. It is the experience of the author that team organization allows easier coordination of activities and better learning. Nevertheless, when dividing the work among teams, it is necessary to stress the importance of teams being consistent in the way they perform their setups and collect their data.

The number of teams to form for an investigation (and thereby, the number of students per team) should be decided before you begin an investigation. Some of the investigations have been written to employ a specific number of teams (often three or four), while in others, the number of teams is more flexible. But even for investigations that specify a number of teams, the investigation can always be modified to use some other number. The main consideration in deciding on the right number of teams is the number of students in the class or section, but another consideration in some cases is the number of agroecosystems (or other such units) available for study.

Because of the central role of teams, the investigations are written for single teams; that is, the procedures describe what a single team does, not what the class or an individual does.

Using the Datasheets

Most investigations include at least one datasheet. In addition to serving as a place in which to record data, a datasheet helps provide an organized picture of what is being analyzed or compared. To extend this latter purpose, datasheets containing example data are provided within each investigation, along with blank datasheets.

For some investigations, more than one copy of a datasheet is needed for each student; in these cases, this need is indicated by a line in the "Materials, Equipment, and Facilities" section, and the extra copies must be photocopied. It is important that the instructor keep a clean copy of the lab manual as a source of photocopy-able datasheets and worksheets.

The exact function of each datasheet depends on the investigation: in some cases, they serve as a central place for all of a team's data, and sometimes even a whole class's data. In other cases, students record "raw" data in their lab notebooks and use the datasheet for "processed" data only. Some investigations also include worksheets, which students can use for making calculations based on collected data.

Sharing Data among Teams

Although each team's dataset is typically adequate on its own for a basis for analysis, most investigations are designed so that the most interesting and instructive analysis and comparison occurs only after teams have shared their data and all teams have available a class or section dataset. There are a number of ways for instructors or TAs to facilitate the sharing of data.

- Teams can be instructed to post copies of their datasheets.
- A *master* datasheet containing the data from all the teams can be kept on a chalkboard or a transparency (which is displayed when appropriate).
- Data can be collected in an electronic spreadsheet, which can then be distributed by printed handouts, by e-mail, or on a website.

The latter method, using an electronic spreadsheet, has a number of distinct advantages (including ease of editing, calculation of means and statistics, and distribution) but requires that one person (typically the instructor or TA) take responsibility for its construction and maintenance.

Assigning Work

Students should be expected to write a lab report for every investigation carried out during the course. Writing a lab report is an essential part of the learning process and an excellent means of evaluating students' performance, participation, and understanding of the work. Each investigation includes suggestions for what to include in the write-up, and general expectations for lab report format and scope are described in the introduction to this manual.

Lab reports can either be done individually or as a team. If you assign team reports, be sure to stress the importance of collaboration throughout the data analysis and writing processes and of equal division of work. If you assign individual reports, make it clear that teamwork is still important during the writing phase.

Students should also be expected to keep a lab notebook during the course. (The purpose of a lab notebook is discussed in the Introduction.) Students will find that keeping careful notes about procedures, observations, and data gathering will be invaluable when it comes time to writing the lab report. Instructors and/or TAs may want to have students turn in their lab notebooks at various times during the course so that the notebooks may be checked and constructively critiqued, and perhaps evaluated.

Section VII

General Appendices

Appendix B: Measurement Equivalents

Mass

1 ounce (oz) = 28.349527 g = 0.0625 lb

1 pound (lb) = 16 oz = 453.59243 g

1 ton (US) = 2000 lb = 907.18486 kg = 32,000 oz

1 milligram (mg) = 0.001 g

1 gram (g) = 1000 mg = 0.035274 oz = 0.0022046 lb = 0.001 kg

1 kilogram (kg) = 1000 g = 35.273957 oz = 2.20462 lb

1 ton (metric) = 2204.6 lb = 1000 kg = 1.1 ton (US)

Area

1 square inch (in.^{2}) = 6.451626 cm^2 = 0.0069444 ft^2

1 square foot (ft^2) = 144 in.^{2} = 0.111 yd^2 = 0.0929 m^2

1 square yard (yd^2) = 9 ft^2 = 1296 in.^{2} = 0.83613 m^2

1 acre (ac) = 43,560 ft^2 = 4,840 yd^2 = 4046.873 m^2 = 0.404687 ha = 0.0015625 mi^2

1 square mile (mi^2) = 640 ac = 1 section = 258.9998 ha = 3,097,600 yd^2 = 2,589,998 km^2

1 square millimeter (mm^2) = 0.01 cm^2 = 0.000001 m^2 = 0.00155 in.^{2}

1 square centimeter (cm^2) = 100 mm^2 = 0.155 in.^{2} = 0.001076 ft^2

1 square meter (m^2) = 10.76387 ft^2 = 1550 in.^{2} = 1.195985 yd^2 = 1,000,000 mm^2 = 10,000 cm^2

1 hectare (ha) = 2.471 ac = 395.367 rod^2 = 10,000 m^2 = 0.01 km^2 = 0.0039 mi^2

1 square kilometer (km^2) = 0.3861 mi^2 = 247.1 ac = 100 ha = 1,000,000 m^2

Volume

1 cubic inch (in.^{3}) = 16.3872 cm^3 = 0.00433 US gal

1 cubic foot (ft^3) = 0.80356 bu = 1728 in.^{3} = 0.037037 yd^3 = 0.028317 m^3 = 7.4805 US gal = 6.229 imperial gal = 28.316 L = 29.922 qt (liquid) = 25.714 qt (dry)

1 cubic foot (ft^3) of dry soil (approximate) = 90 lb (sandy), 80 lb (loamy), 75 lb (clay)

1 bushel (bu) dry soil (approximate) = 112 lb (sandy), 100 lb (loamy), 94 lb (clay)

1 cubic yard (yd^3) = 27 ft^3 = 46,656 in.^{3} = 764.559 L = 202 US gal = 0.764559 m^3

1 cubic centimeter (cm^3, cc) = 0.06102 in.^{3} = 1000 mm^3 = 0.000001 m^3

1 cubic meter (m^3) = 1.30794 yd^3 = 35.3144 ft^3 = 28.377 bu = 264.173 US gal = 61,023 in.^{3} = 1,000,000 cm^3

1 dry quart (qt) = 67.2 in.^{3} = 1.1012 L = 0.125 pk = 0.03125 bu = 0.038889 ft^3 = 67.2 in.^{3}

1 peck (pk) = 0.25 bu = 2 gal = 8 qt (dry) = 8.80958 L = 537.605 in.^{3}

1 bushel (bu) = 4 pk = 32 qt (dry) = 1.2445 ft^3 = 35.2383 L = 2150.42 in.^{3}

1 fluid ounce (fl oz) = 2 tablespoons = 0.125 c = 0.03125 qt (liquid) = 29.573 mL = 1.80469 in.^{3} = 0.029573 L

1 liquid quart (qt) = 32 fl oz = 57.749 in.^{3} = 0.25 US gal = 0.946333 L = 0.3342 ft^3

1 US gallon (gal) = 4 qt = 128 fl oz = 0.1337 ft^3 = 3785.4 mL = 231 in.^{3} = 8.337 lb water = 3.782 kg water

1 milliliter (mL) = 1 cm^3 = 0.001 L = 0.061 in.^{3} = 0.03815 fl oz

1 liter (L) = 1000 mL = 1.0567 liquid qt = 0.9081 dry qt = 0.264178 US gal = 33.8147 fl oz = 61.03 in.^{3} = 0.0353 ft^3 =
      0.02838 bu = 0.001308 yd^3

1 acre-inch of water = 27,152.4 gal

Distance and Length

1 inch (in.) = 25.4 mm = 2.54 cm = 0.0254 m = 0.083333 ft

1 foot (ft) = 12 in. = 0.3333 yd = 0.060606 rod = 30.48 cm = 0.3048 m

1 yard (yd) = 36 in. = 0.9144 m

1 mile (mi) = 5280 ft = 1,609.35 m = 63,360 in.

1 millimeter (mm) = 0.1 cm = 0.001 m = 1000 µm = 0.03937 in.

1 centimeter (cm) = 10 mm = 0.01 m = 0.3937 in. = 0.0328 ft

1 meter (m) = 100 cm = 39.37 in. = 3.2808 ft = 1.0936 yd

1 kilometer (km) = 3280.8 ft = 1093.6 yd = 0.62137 mi

1 nautical mile = 1.85 km = 1.15 mi

Energy and Power

1 calorie (cal) = 4.184 joules (J)

1 British thermal unit (Btu) = 252 cal

1 kilocalorie (kcal) = 1000 cal = 3.97 Btu = 0.00116 kWh

1 joule (J) = 0.239 cal = 107 ergs

1 kilojoule (kJ) = 1000 J = 0.949 Btu = 0.000278 kWh

1 kilowatt-hour (kWh) = 860 kcal = 3400 Btu

1 watt (W) = 1 joule per second

1 horsepower (hp) = 746 W = 2542 Btu per hour

Temperature

°C = (°F − 32.0) ÷ 1.80

°F = (°C × 1.80) + 32.0

Dilutions

1 part per million (ppm) = 1 mg per L = 1 mg per kg = 0.0001% = 0.013 oz by weight in 100 gal = 0.379 g in 100 gal

1 percent (%) = 10,000 ppm = 10 g per L = 1.28 oz by weight per gal

Appendix C: Material and Equipment Suppliers

The author has found the following suppliers to be useful sources of seeds, tools, equipment, and gardening supplies. There are many other reputable suppliers not included here.

Edmund Scientific
532 Main Street, Tonawanda, NY 14150, 800-818-4955
www.scientificsonline.com
> Supplier of scientific materials, equipment, and supplies for education and research.

Forestry Suppliers, Inc.
P.O. Box 8397, Jackson, MS 39284, 800-647-5368
www.forestry-suppliers.com
> A leading supplier of products for the broad field of natural resource management, including environmental science and agriculture. A nice assortment of sampling, measuring, and testing kits and tools.

Gardens Alive
5100 Schenley Place, Lawrenceburg, IN 47025, 513-354-1482
www.gardensalive.com
> Carries a wide selection of environmentally responsible pest-control products, beneficial insects, composting products, soil amendments, and gardening supplies.

Johnny's Selected Seed
955 Benton Avenue, Winslow, ME 04901, 877-564-6697
www.johnnyseeds.com
> An excellent source of heritage, small-farm, and garden-adapted seed, with a fine choice of Oriental, European, and American vegetables. Most varieties for cooler, shorter-season climates, with most usually available as untreated seed.

LaMotte Company
802 Washington Avenue, Chestertown, MD 21620, 800-344-3100
www.lamotte.com
> Makers of soil and water testing equipment and supplies and outdoor monitoring equipment.

Peaceful Valley Farm Supply
P.O. Box 2209, Grass Valley, CA 95945, 888-784-1722
www.groworganic.com
    Extensive catalog of organic gardening and farming supplies, including fertilizers, beneficial insects,
    tools, seeds, and monitoring equipment.

Shepherd's Garden Seeds
6060 Graham Hill Road, Felton, CA 95018, 888-880-7228
www.reneesgarden.com
    Seeds specifically selected for kitchen gardens. A marvelous assortment of gourmet vegetables, herbs,
    and flowers.

Territorial Seed Company
PO Box 158, Cottage Grove, OR 97424, 800-626-0866
www.territorialseed.com
    Offers seeds and plants selected for optimal growth in the maritime northwest of North America
    (Northern California, Oregon, Washington, and British Columbia).

GREYSCALE

BIN TRAVELER FORM

Cut By _Sordw Sorjan_   Qty _90_   Date _2/16/2024_

Scanned By _____   Qty _____   Date _____

Scanned Batch IDs

_____   _____   _____

Notes / Exception

_____

_____

_____

_____

_____